Our Robots, Ourselves

ALSO BY DAVID A. MINDELL

Digital Apollo:
Human and Machine in Spaceflight

Iron Coffin:
War, Technology, and Experience Aboard the USS Monitor

Between Human and Machine:
Feedback, Control, and Computing Before Cybernetics

Our Robots, Ourselves

ROBOTICS AND THE MYTHS OF AUTONOMY

DAVID A. MINDELL

VIKING

VIKING

An imprint of Penguin Random House LLC

375 Hudson Street

New York, New York 10014

penguin.com

ISBN 978-0-525-42697-4

Printed in the United States of America

1   3   5   7   9   10   8   6   4   2

Set in Scala OT with DIN Next LT Pro

Designed by Daniel Lagin

Dedicated to the memory of

Martin Bowen, early robotic explorer

and

Seth Teller, humane roboticist

Poetry is the journal of a sea animal living on land, wanting to fly in the air.

—CARL SANDBURG

CONTENTS

Our Robots, Ourselves

CHAPTER 1

Human, Remote, Autonomous

LATE IN THE NIGHT, HIGH ABOVE THE ATLANTIC OCEAN IN THE long, open stretch between Brazil and Africa, an airliner encountered rough weather. Ice clogged the small tubes on the aircraft's nose that detected airspeed and transmitted the data to the computers flying the plane. The computers could have continued flying without the information, but they had been told by their programmers that they could not.

The automated, fly-by-wire system gave up, turned itself off, and handed control to the human pilots in the cockpit: thirty-two-year-old Pierre Cedric Bonin and thirty-seven-year-old David Robert. Bonin and Robert, both relaxed and a little fatigued, were caught by surprise, suddenly responsible for hand flying a large airliner at high altitude in bad weather at night. It is a challenging task under the best of circumstances, and one they had not handled recently. Their captain, fifty-eight-year-old Marc Debois, was off duty back in the cabin. They had to waste precious attention to summon him.

Even though the aircraft was flying straight and level when the computers tripped off, the pilots struggled to make sense of the bad air data. One man pulled back, the other pushed forward on his control stick. They continued straight and level for about a minute, then lost control.

On June 1, 2009, Air France flight 447 spiraled into the ocean, killing more than two hundred passengers and crew. It disappeared below the waves, nearly without a trace.

In the global, interconnected system of international aviation, it is unacceptable for an airliner to simply disappear. A massive, coordinated search followed. In just a few days traces of flight 447 were located on the ocean's surface. Finding the bulk of the wreckage, however, and the black box data recorders that held the keys to the accident's causes, required hunting across a vast seafloor, and proved frustratingly slow.

More than two years later, two miles deep on the seafloor, nearly beneath the very spot where the airliner hit the ocean, an autonomous underwater vehicle, or AUV, called Remus 6000 glided quietly through the darkness and extreme pressure. Moving at just faster than a human walking pace, the torpedo-shaped robot maintained a precise altitude of about two hundred feet off the bottom, a position at which its ultrasonic scanning sonar returns the sharpest images. As the sonars pinged to about a half mile out either side, the robot collected gigabytes of data from the echoes.

The terrain is mountainous, so the seafloor rose quickly. Despite its intelligence, the robot occasionally bumped into the bottom, mostly without injury. Three such robots worked in a coordinated dance: two searched underwater at any given time, while a third one rested on a surface ship in a three-hour pit stop with its human handlers to offload data, charge batteries, and take on new search plans.

On the ship, a team of twelve engineers from the Woods Hole Oceanographic Institution, including leader Mike Purcell, who spearheaded the design and development of the searching vehicles, worked in twelve-hour shifts, busy as any pit crew. When a vehicle came to the surface, it took about forty-five minutes for the engineers to download the data it collected into a computer, then an additional half hour to process those data to enable a quick, preliminary scroll-through on a screen.

Looking over their shoulders were French and German investigators, and representatives from Air France. The mood was calculating and deliberate, but tense: the stakes were high for French national pride, for the airliner's manufacturer, Airbus, and for the safety of all air travel. Several prior expeditions had tried and failed. In France, Brazil, and around the world, families awaited word.

Interpreting sonar data requires subtle judgment not easily left solely to a computer. Purcell and his engineers relied on years of experience. On their screens, they reviewed miles and miles of rocky reflections alternating with smooth bottom. The pattern went on for five days before the monotony broke: a crowd of fragments appeared, then a debris field—a strong signal of human-made artifacts in the ocean desert. Suggestive, but still not definitive.

The engineers reprogrammed the vehicles to return to the debris and "fly" back and forth across it, this time close enough that onboard lights and cameras could take pictures from about thirty feet off the bottom. When the vehicles brought the images back to the surface, engineers and investigators recognized the debris and had their answer: they had found the wreckage of flight 447, gravesite of hundreds.

Soon, another team returned with a different kind of robot, a remotely operated vehicle (ROV), a heavy-lift vehicle specially designed for deep salvage, connected by a cable to the ship. Using the maps created by the successful search, the ROV located the airliner's black box voice and data recorders and brought them to the surface. The doomed pilots' last minutes were recovered from the ocean, and investigators could now reconstruct the fatal confusion aboard the automated airliner. The ROV then set about the grim task of retrieving human remains.

The Air France 447 crash and recovery linked advanced automation and robotics across two extreme environments: the high atmosphere and the deep sea. The aircraft plunged into the ocean because of failures in human interaction with automated systems; the wreckage was then discovered by humans operating remote and autonomous robots.

While the words (and their commonly perceived meanings) suggest that automated and autonomous systems are self-acting, in both cases the failure or success of the systems derived not from the machines or the humans operating on their own, but from people and machines operating together. Human pilots struggled to fly an aircraft that had been automated for greater safety and reliability; networks of ships, satellites, and floating buoys helped pinpoint locations; engineers interpreted and acted on data produced by robots. Automated and autonomous vehicles constantly returned to their human makers for information, energy, and guidance.

Air France 447 made tragically clear that as we constantly adapt to and reshape our surroundings, we are also remaking ourselves. How could pilots have become so dependent on computers that they flew a perfectly good airliner into the sea? What becomes of the human roles in activities like transportation, exploration, and warfare when more and more of the critical tasks seem to be done by machines?

In the extreme view, some believe that humans are about to become obsolete, that robots are "only one software upgrade away" from full autonomy, as *Scientific American* has recently argued. And they tell us that the robots are coming—coming to more familiar environments. A new concern for the strange and uncertain potentials of artificial intelligence has arisen out of claims that we are on the cusp of superintelligence. Our world is about to be transformed, indeed is already being transformed, by robotics and automation. Start-ups are popping up, drawing on old dreams of smart machines to help us with professional duties, physical labor, and the mundane tasks of daily life. Robots living and working alongside humans in physical, cognitive, and emotional intimacy have emerged as a growing and promising subject of research. *Autonomy*—the dream that robots will one day act as fully independent agents—remains a source of inspiration, innovation, and concern.

The excitement is in the thrill of experimentation; the precise forms of these technologies are far from certain, much less their social, psychological, and cognitive implications. How will our robots change us? In whose image will we make them? In the domain of work, what will become of our traditional roles—scientist, lawyer, doctor, soldier, manager, even driver and sweeper—when the tasks are altered by machines? How will we live and work?

We need not speculate: much of this future is with us today, if not in daily life then in extreme environments, where we have been using robotics and automation for decades. In the high atmosphere, the deep ocean, and outer space humans cannot exist on their own. The demands of placing human beings in these dangerous settings have forced the people who work in them to build and adopt robotics and automation earlier than those in other, more familiar realms.

Extreme environments press the relationships between people and machines to their limits. They have long been sites of innovation. Here engineers have the freest hand to experiment. Despite the physical isolation, here the technologies' cognitive and social effects first become apparent. Because human lives, expensive equipment, and important missions are at stake, autonomy must always be tempered with safety and reliability.

In these environments, the mess and busyness of daily life are temporarily suspended, and we find, set off from the surrounding darkness, brief, dream-like allegories of human life and technology. The social and technological forces at work on an airliner's flight deck, or inside a deep-diving submersible, are not fundamentally different from those in a factory, an office, or an automobile. But in extreme environments they appear in condensed, intense form, and are hence easier to grasp. Every airplane flight is a story, and so is every oceanographic expedition, every space flight, every military mission. Through these stories of specific people and machines we can glean subtle, emerging dynamics.

Extreme environments teach us about our near future, when similar technologies might pervade automobiles, health care, education, and other human endeavors. Human-operated, remotely controlled, and autonomous vehicles represent the leading edge of machine and human potential, new forms of presence and experience, while drawing our attention to the perils, ethical implications, and unintended consequences of living with smart machines. We see a future where human knowledge and presence will be more crucial than ever, if in some ways strange and unfamiliar.

And these machines are just cool. I'm not alone in my lifelong fascination with airplanes, spacecraft, and submarines. Indeed, technological enthusiasm, as much as the search for practical utility, drives the stories that follow. It's no coincidence that similar stories are so often the subject of science fiction—something about people and machines at the limits of their abilities captures the imagination, engages our wonder, and stirs hopes about who we might become.

This enthusiasm sometimes reflects a naïve faith in the promise of technology. But when mature it is an enthusiasm for basic philosophical and humanistic questions: Who are we? How do we relate to our work and to one another? How do our creations expand our experience? How can we best live in an uncertain world? These questions lurk barely below the surface as we talk to people who build and operate robots and vehicles.

Join me as I draw on firsthand experience, extensive interviews, and the latest research from MIT and elsewhere to explore experiences of robotics and automation in the extreme environments of the deep ocean and in aviation (civil and military) and spaceflight. It is not an imagination of the future, but a picture of today: we'll see how people operate with and through robots and autonomy and how their interactions affect their work, their experiences, and their skills and knowledge.

Our stories begin where I began, in the deep ocean. Twenty-five years ago, as an engineer designing embedded computers and instruments for deep-ocean robots, I was surprised to find that technologies were changing in unexpected ways the work of oceanography, the ways of doing science, the meaning of being an oceanographer.

The realization led to two parallel careers. As a scholar, I study the human implications of machinery, from ironclad warships in the American Civil War to the computers and software that helped the Apollo astronauts land on the moon. As an engineer, I bring that research to bear on present-day projects, building robots and vehicles designed to work in intimate partnership with people. In the stories that follow I appear in some as a participant, in others as an observer, and in still others as both.

These years of experience, research, and conversation have convinced me that we need to change the way we think about robots. The language we use for them is more often from twentieth-century science fiction than from the technological lives we lead today. Remotely piloted aircraft, for example, are referred to as "drones," as though they were mindless automata, when actually they are tightly controlled by people. Robots are imagined (and sold) as fully autonomous agents, when even today's modest autonomy is shot through with human imagination. Rather than being threatening automata, the robots we use so variously are embedded, as are we, in social and technical networks. In the pages ahead, we will explore many examples of how we work together with our machines. It's the combinations that matter.

It is time to review what the robots of today actually do, to deepen our understanding of our relationships with these often astoundingly capable human creations. I argue for a deeply researched empirical conclusion: whatever they might do in a laboratory, as robots move closer to environments with human lives and real resources at stake, we tend to add more human approvals and interventions to govern their autonomy. My argument here is not that machines are not intelligent, nor that someday they might not be. Rather, my argument is that such machines are not *inhuman*.

Let us name three mythologies of twentieth-century robotics and automation. First, there is *the myth of linear progress*, the idea that technology evolves from direct human involvement to remote presence and then to fully autonomous robots. Political scientist Peter W. Singer, a prominent public advocate for autonomous systems, epitomizes this mythology when he writes that "this concept of keeping the human in the loop is already being eroded by both policymakers and the technology itself, which are both rapidly moving toward pushing humans out of the loop."

Yet there is no evidence to suggest that this is a natural evolution, that the "technology itself," as Singer puts it, does any such thing. In fact there is good evidence that people are moving into deeper intimacy with their machines.

We repeatedly find human, remote, and autonomous vehicles evolving together, each affecting the other. Unmanned aircraft, for example, cannot occupy the national airspace without the task of piloting manned aircraft changing too. In another realm, new robotic techniques for servicing spacecraft changed the way human astronauts serviced the Hubble Space Telescope. The most advanced (and difficult) technologies are not those that stand apart from people, but those that are most deeply embedded in, and responsive to, human and social networks.

Second is *the myth of replacement*, the idea that machines take over human jobs, one for one. This myth is a twentieth-century version of what I call the iron horse phenomenon. Railroads were initially imagined to replace horses, but trains proved to be very poor horses. Railroads came into their own when people learned to do entirely new things with them. Human-factors researchers and cognitive scientists find that rarely does automation simply "mechanize" a human task; rather, it tends to make the task more complex, often increasing the workload (or shifting it around). Remotely piloted aircraft do not replicate the missions that manned aircraft carry out; they do new things. Remote robots on Mars do not copy human field science; they and their human partners learn to do a new kind of remote, robotic field science.

Finally, we have *the myth of full autonomy*, the utopian idea that robots, today or in the future, can operate entirely on their own. Yes, automation can certainly take on parts of tasks previously accomplished by humans, and machines do act on their own in response to their environments for certain periods of time. But the machine that operates entirely independently of human direction is a useless machine. Only a rock is truly autonomous (and even a rock was formed and placed by its environment). Automation changes the type of human involvement required and transforms but does not eliminate it. For any apparently autonomous system, we can always find the wrapper of human control that makes it useful and returns meaningful data. In the words of a recent report by the Defense Science Board, "there are no fully autonomous systems just as there are no fully autonomous soldiers, sailors, airmen or Marines."

To move our notions of robotics and automation, and particularly the newer idea of autonomy, into the twenty-first century, we must deeply grasp how human intentions, plans, and assumptions are always built into machines. Every operator, when controlling his or her machine, interacts with designers and programmers who are still present inside it—perhaps through design and coding done many years before. The computers on Air France 447 could have continued to fly the plane even without input from the faulty airspeed data, but they were programmed by people not to. Even if software takes actions that could not have been predicted, it acts within frames and constraints imposed upon it by its creators. How a system is designed, by whom, and for what purpose shapes its abilities and its relationships with the people who use it.

My goal is to move beyond these myths and toward a vision of situated autonomy for the twenty-first century. Through the stories that follow, I aim to redefine the public conversation and provide a conceptual map for a new era.

As the basis for that map, I will rely throughout the book on *human, remote,* and *autonomous* when referring to vehicles and robots. The first substitutes for the awkward "manned," so you can read "human" as shorthand for "human occupied." These are of course old and familiar types of vehicles like ships, aircraft, trains, and automobiles, in which peoples' bodies travel with the machines. People generally do not consider human-occupied systems to be robots at all, although they do increasingly resemble robots that people sit inside.

"Remote," as in remotely operated vehicles (ROVs), simply makes a statement about where the operator's body is, in relation to the vehicle. Yet even when the cognitive task of operating a remote system is nearly identical to that of a direct physical operator, great cultural weight is attached to the presence or absence of the body, and the risks it might undergo. In the most salient example, remotely fighting a war from thousands of miles away is a different experience from traditional soldiering. As a cognitive phenomenon, human presence is intertwined with social relationships.

Automation is also a twentieth-century idea, and still carries a mechanical sense of machines that step through predefined procedures; "automated" is the term commonly used to describe the computers on airliners, even though they contain modern, sophisticated algorithms. "Autonomy" is the more current buzzword, one that describes one of the top priorities of research for a shrinking Department of Defense. Some clearly distinguish autonomy from automation, but I see the difference as a matter of degree, where autonomy connotes a broader sense of self-determination than simple feedback loops and incorporates a panoply of ideas imported from artificial intelligence and other disciplines. And of course the idea of the autonomy of individuals and groups pervades current debates in politics, philosophy, medicine, and sociology. This should come as no surprise, as technologists often borrow social ideas to describe their machines.

Even within engineering, autonomy means several different things. Autonomy in spacecraft design refers to the onboard processing that takes care of the vehicle (whether an orbiting probe or a mobile robot) as distinct from tasks like mission planning. At the Massachusetts Institute of Technology, where I teach, the curriculum of engineering courses on autonomy covers mostly "path planning"—how to get from here to there in a reasonable amount of time without hitting anything. In other settings autonomy is analogous to intelligence, the ability to make human-like decisions about tasks and situations, or the ability to do things beyond what designers intended or foresaw. Autonomous underwater vehicles (AUVs) are so named because they are untethered, and contrast with remotely operated vehicles (ROVs), which are connected by long cables. Yet AUV engineers recognize that their vehicles are only semiautonomous, as they are only sometimes fully out of touch.

The term "autonomous" allows a great deal of leeway; it describes how a vehicle is controlled, which may well change from moment to moment. One recent report introduces the term "increasing autonomy" to describe its essentially relative nature, and to emphasize how "full" autonomy—describing machines that require no human input—will always be out of reach. For our purposes, a working definition of autonomy is: a human-designed means for transforming data sensed from the environment into purposeful plans and actions.

Language matters, and it colors debates. But we need not get stuck on it; I will often rely on the language (which is sometimes imprecise) used by the people I study. The weight of this book rests not on definitions but on stories of work: How are people using these systems in the real world, experiencing, exploring, even fighting and killing? What are they actually doing?

Focusing on lived experiences of designers and users helps clarify the debates. For example, the word "drone" obscures the essentially human nature of the robots and attributes their ill effects to abstract ideas like "technology" or "automation." When we visit the Predator operators' intimate lairs we will discover that they are not conducting automated warfare—people are still inventing, programming, and operating machines. Much remains to debate about the ethics and policy of remote assassinations carried out by unmanned aircraft with remote operators, or the privacy concerns with similar devices operating in domestic airspace. But these debates are about the nature, location, and timing of *human* decisions and actions, not about machines that operate autonomously.

Hence the issues are not manned versus unmanned, nor human-controlled versus autonomous. The questions at the heart of this book are: *Where are the people? Which people are they? What are they doing? When are they doing it?*

Where are the people? (On a ship . . . in the air . . . inside the machine . . . in an office?)

The operator of the Predator drone may be doing something very similar to the pilot of an aircraft—monitoring onboard systems, absorbing data, making decisions, and taking actions. But his or her body is in a different place, perhaps even several thousand miles away from the results of the work. This difference matters. The task is different. The risks are different, as are the politics.

People's minds can travel to other places, other countries, other planets. Knowledge through the mind and senses is one kind of knowledge, and knowledge through the body (where you eat, sleep, socialize, and defecate) is another. Which one we privilege at any given time has consequences for those involved.

Which people are they? (Pilots . . . engineers . . . scientists . . . unskilled workers . . . managers?)

Change the technology and you change the task, and you change the nature of the worker—in fact you change the entire population of people who can operate a system. Becoming an air force pilot takes years of training, and places one at the top of the labor hierarchy. Does operating a remote aircraft require the same skills and traits of character? From which social classes does the task draw its workforce? The rise of automation in commercial-airline cockpits has corresponded to the expanding demographics of the pilot population, both within industrialized countries and around the globe. Is an explorer someone who travels into a dangerous environment, or someone who sits at home behind a computer? Do you have to like living on board a ship to be an oceanographer? Can you explore Mars if you're confined to a wheelchair? Who are the new pilots, explorers, and scientists who work through remote data?

What are they doing? (Flying . . . operating . . . interpreting data . . . communicating?)

A physical task becomes a visual display, and then a cognitive task. What once required strength now requires attention, patience, quick reactions. Is a pilot mainly moving her hands on the controls to fly the aircraft? Or is she punching key commands into an autopilot or flight computer to program the craft's trajectory? Where exactly is the human judgment she is adding? What is the role of the engineer who programmed her computer, or the airline technician who set it up?

When are they doing it? (In real time . . . after some delay . . . months or years earlier?)

Flying a traditional airplane takes place in real time—the human inputs come as the events are happening and have immediate results. In a spaceflight scenario, the vehicle might be on Mars (or approaching a distant asteroid), in which case it might take twenty minutes for the vehicle to receive the command, and twenty minutes for the operator to see that the action has occurred. Or we might say that craft is landing "automatically," when actually we can think of it as landing under the control of the programmers who gave it instructions months or years earlier (although we may need to update our notions of "control"). Operating an automated system can be like cooperating with a ghost.

These simple questions draw our attention to shifts and reorientations. New forms of human presence and action are not trivial, nor are they equivalent—a pilot who risks bodily harm above the battlefield has a different cultural identity from one who operates from a remote ground-control station. But the changes are also surprising—the remote operator may feel more present on the battlefield than pilots flying high above it. The scientific data extracted from the moon may be the same, or better, when collected by a remote rover than by a human who is physically present in the environment. But the cultural experience of lunar exploration is different from being there.

Let's replace dated mythologies with rich, human pictures of how we actually build and operate robots and automated systems in the real world. The stories that follow are at once technological and humanistic. We shall see human, remote, and autonomous machines as ways to move and reorient human presence and action in time and in space. The essence of the book boils down to this: it is not "manned" versus "unmanned" that matters, but rather, where are the people? Which people? What are they doing? And when?

The last, and most difficult questions, then, are:

How does human experience change? And why does it matter?

the hatch behind me. The sky disappeared with a feeling of finality; I would not get out for days. I stood aside as the crew prepared to dive. Checks, calls, communications; in a flurry of hand-cranked valves the sub began to descend with a gentle downhill pitch.

My bunk was on top of the narrow corridor. Surrounded by pipes and brackets, it had only a small opening at the foot end. From there I had to wriggle inside to get my head in place for sleeping. Once in position, I could not turn over. Lying on my back, a bunch of pipes hung right in front of my face, and a few inches behind them was the sub's steel hull. On the other side of that was, well, three thousand feet of water. The first time I slept up there I awoke feeling claustrophobic and had to get down immediately and walk around to relax. The second time, it seemed a little more cozy but still made me nervous. By night three it felt like home.

After descending for a few minutes we reached the bottom just outside Skerki Bank, about three thousand feet deep, and began our survey, looking for telltale signs of shipwrecks. On its sides, NR-1 had "side scan" sonars, which could see a few hundred meters out to each side. But NR-1's main feature was the forward-looking sonar. Every couple of seconds, the sonar on the nose of the submarine assaulted the watery space with a ping of high-frequency sound, then collected the echoes and displayed them on a computer screen. The sonar was designed to look up under ice for possible Russian submarines. Mounted on the NR-1, it pointed downward and forward of the sub; it could see a soda can from three thousand yards (and we saw quite a few on the bottom of the Mediterranean).

The trouble was, the sonar only displayed "targets"—fuzzy blobs of pixels. To make out what they were, the crew had to laboriously drive over to each target and closely observe it either out the window or with

NR-1's many cameras. And NR-1 was quite slow; it could make just a knot or two across the bottom, about a human walking pace. If we detected something three thousand yards out on the sonar, it could take hours to drive over to it and have a look.

About an hour after we began our dive, Scott, the lieutenant on navigation watch who also served as the sonar man, noticed something on the display. It was a target not more than a few pixels across, but Scott thought it might be man-made. It had a denser inner area surrounded, halo-like, by a ring of less dense reflections. This is not what rocks look like on sonar. As we moved past the target, the position and appearance of the blob did not change, though the grazing angle of the sonar changed—another indication of something solid, substantial, and possibly of human origin. Scott recommended that we depart from our track line and approach.

It was the start of what was supposed to be a two-day survey and already my leadership was being tested. Just an hour or two had passed since Ballard's admonition to stick to the track lines. But I had to trust the crew. If this departure turned out to be a wild-goose chase, then I'd earn the credibility to turn down future requests.

I descended into the NR-1's viewing area, a cramped compartment on the bottom of the hull with small windows. We were traveling about forty feet off the seafloor at a leisurely pace. Outside I saw undifferentiated green, the color a result of NR-1's green lights. Squinting, I could make out the sand, and get a sense of motion only when a ripple or rock slid by to break the visual monotony. As we approached the mysterious target, I prepared to see a pile of rocks.

Instead, what emerged out of the green filled me with awe. Ceramic jars from an ancient world, more than a hundred of them, lay about the ocean floor. They were scattered, but in an identifiable pattern in two distinct piles about ten meters apart. This was the site of an ancient shipwreck. Long ago, the wooden hull rotted away, leaving the cargo exposed just as it was stacked in the hull. Lead stocks from two lead anchors clearly identified the bow. The wreck was pristine, untouched and unseen since settling here more than two thousand years ago. As the first person to see it since the day it sank, I was moved by the magnitude of the passage of time, and by the power of physical presence to abridge that time.

I named our discovery Skerki D, a scientific-sounding way to describe the fourth known shipwreck on Skerki Bank. We announced it to our colleagues on the surface through an underwater telephone, a

The U.S. Navy's NR-1 nuclear research submarine above Skerki D, the remains of a first-century BCE shipwreck 800 meters (3,000 feet) deep in the Mediterranean Sea.

(COURTESY NATIONAL GEOGRAPHIC SOCIETY)

scratchy, uneven channel that, at its best, garbled voices like an old walkie-talkie. We carefully noted the position and took a lot of pictures.

At the end of our survey, after about a day and a half, we planned to return to the surface. Formal, clear language squeaked over the underwater telephone: "Interrogative: what is weather on surface?" A gale was brewing up there, which would have made it unsafe for us to transfer back to the *Carolyn Chouest*. So we returned to Skerki D and took a few more pictures. NR-1 has wheels, so we just moved off the site a few hundred feet and planted the submarine on the bottom. There we sat, at three thousand feet, waiting for the weather to clear up—for nearly two days, watching war movies in the tiny galley.

Finally we got word that the weather was clearing, and we ascended as quickly as we had dived.

I returned to the *Carolyn Chouest* feeling serene but excited about our successful hunting, only to find my shipboard colleagues green, a little seasick, and tired from a rough couple of days. We had indeed been in a different world, less than a mile away but straight down.

What came next was a natural experiment comparing the emotional power of embodied experience to the cognitive power of remote presence. For I was not a native submariner but a robotics engineer. The amount of time I spent physically on the seafloor was dwarfed by the amount of time I spent remotely there, telepresent through the medium of remote robots and fiber-optic cables.

My home technology was the remote robot *Jason*, built and run by the Deep Submergence Laboratory of the Woods Hole Oceanographic Institution (WHOI). The Volkswagen-sized *Jason* waited out the rough weather lashed to the deck of the *Carolyn Chouest*. As soon as the weather cleared and NR-1 got out of the way, we quickly tasked *Jason* to carry out an intense, computer-controlled survey of the wreck site.

We sat on board the ship in a darkened, air-conditioned control room while *Jason*, connected to the surface via a high-bandwidth fiber-optic cable, descended the depths to the Skerki D site. We watched on video, monitored sensors, and frenetically programmed computers. On this particular dive, seven years of work came together: sensors, precision navigation systems, and computerized controls coordinated to hover *Jason* above Skerki D and move it at a snail's pace to run precise track lines, just a meter apart, above the wreck site. Sonars and digital cameras bounced sound and light off the wreck, gathering gobs of data and transmitting it to hard drives on the ship. An acoustic navigation system I had built monitored *Jason*'s position with subcentimeter precision several times per second, giving exact location tags to all the data.

Then engineers and graduate students set to work, compiling the images into extensive photomosaics of the site and assembling the sonar data into a high-precision topographical map. This map connected navigation, computers, sensors, and data processing into a single pipeline. We had done pieces of this before, but never all together, and never on such an interesting and important site.

The robotic survey expanded and quantified what I had seen out the window of NR-1. Where the submarine provided a visceral experience of presence, the robot digitized the seafloor into bits. Then as we pored over the data from the comparative comfort of the surface ship, we explored the virtual site in detail, discovering a great deal about it that was not visible when I was "there."

We could now say the wreck site was about twenty meters long by five meters wide, with two distinct piles of ancient jars called amphoras. Many of the amphoras lay in small craters, apparently scoured out just for them by thousands of years of gentle bottom currents. Most of the amphoras were quite varied in appearance, although three identical ones lay, almost as if they had been lashed together, in a single crater. The seafloor, apparently flat to my naked eye peering through the window, actually had a gentle crescent just a few centimeters high that marked the outline of Skerki D's ship's hull, buried just below the mud line.

When we showed the digital maps to one of the archaeologists on board, he exclaimed, "You've just done in four hours what I spent seven years doing on the last site I excavated." Yet no scuba-diving archaeologist ever had a map nearly as detailed and precise as our map of Skerki D—in fact, it was the most precise map ever made of the ocean floor, albeit of a tiny square in the vast ocean.

The Skerki D survey was the culmination of at least eight years of engineering. We had learned how to digitize the seafloor with ultra-high precision. That would change both what was possible in archaeology and in how we explore human history in the deep sea. We would now learn how to "excavate" an archaeological site without ever touching it. We would now learn how to do a new kind of archaeology focused on the deep water and ancient trade routes that connected civilizations. It would let us ask new questions. But not everyone would welcome the new methods.

Robotic exploration would prove troubling to some and exciting to others. Seeking to understand this resistance led me on a journey of research and discovery spanning twenty years. Before moving forward with that story, however, we need to go back in time to the birth of deep-ocean exploration to see how people came to visit the deep seafloor, and how *Jason* fit into that story.

Today we routinely use robots to explore the ocean. When an airliner goes missing or an oil well bursts, robots are our first—and often our only—eyes on the scene. But as *Jason* developed over time, a passionate debate raged. "Those robots will never be more than engineers' toys," some scientists admonished us. Some actually declared you had to physically visit the seafloor to be a real oceanographer. Prominent archaeologists argued that visiting an ancient shipwreck in the deep ocean with a remote robot, even if solely for the purpose of taking pictures, was inherently unethical.

These archaeologists had developed their methods in shallow water. As scuba divers, they could explore the top few hundred feet of the ocean, which tended to keep them near the coasts. Even most submarines are generally limited to the top few thousand feet or so. They are literally submersible boats designed to travel across the ocean rather than plumb its depths. (Many submarines and robots have been built for shallow, coastal environments, and you can even build one out of parts from a hardware store.)

By contrast, the Deep Submergence Laboratory specialized in very deep water, often several miles deep, which placed us mostly in the mid-oceans or in trenches and subduction zones. These are exotic environments with extreme conditions, high pressures, and other factors that make special demands on machinery and people.

In engineering terms, full ocean depth is 6,000 meters (19,500 feet), which covers more than 90 percent of the seafloor. You need heavy machines to go that deep: to keep a piece of electronics dry at that depth can require huge, cylindrical metal housings that dwarf the circuits they protect in size, weight, and cost. The deepest known part of the ocean, the Mariana Trench, is nearly 11,000 meters (more than 35,000 feet), and requires yet further specialized technology. The mid-ocean ridges, which ring the earth like the seams on a baseball, rise to between 2,000 and 4,000 meters below the surface.

The *Jason* robot was competing not so much with the navy's NR-1 but with the space-age vehicle that Woods Hole had developed in the 1960s to carry scientists' eyes, bodies, and minds directly to these deep spots. Its name was *Alvin*, and for more than two decades this white submersible had been making headlines, collecting scientific data, and capturing the imagination of the public. In this story, the robotic and manned systems evolved together.

The connection between them was Bob Ballard.

Dr. Robert D. Ballard did not invent any of the relevant technologies, nor would he claim to have done so. Trained as a scientist, not an engineer, he's one of those rare scientists who works comfortably with engineers and thinks about doing science with future, possible instruments rather than with those that exist already. Most of the pieces of his robotic systems had been tried in other places. But Ballard had intimate experience with the deep seafloor, developed a vision of remote robotics for that environment, built a laboratory and a team to implement that vision, and led that team on key projects that proved the technological systems. Only now, looking back on his mentorship, do I realize how much his vision of telepresence shaped my own thinking.

Ballard originally came to Woods Hole from California in 1966 as a naval officer during the Vietnam era. His father was an engineer who had worked on inertial guidance systems. Ballard's early jobs were at North American Aviation, studying early submersibles, though he was more interested in the science than the technical details. He began graduate studies in oceanography but was soon called to active duty with the navy. They assigned him as liaison between Woods Hole and the Office of Naval Research (ONR). Although a privately endowed institution like a university, Woods Hole at that time was mostly funded by the navy.

The navy generally had little interest in the deep ocean, which it saw as irrelevant for Cold War combat (combat submarines dive less than two thousand feet). That all changed in 1963, when the nuclear submarine *Thresher* was lost with 129 officers and enlisted men at a depth of 2,600 meters (8,400 feet). The navy realized it had no way to find and recover its own expensive, secret, and dangerous items as they sank to the deep seafloor. Hence ONR began paying for Woods Hole to develop a submersible (distinguished from a submarine by its inability to travel across the ocean under its own power; a submersible is carried to a site by a mother ship, and its dive profile is primarily vertical). It consisted of a precisely shaped sphere of specialized HY-100 steel (later titanium), seven feet in diameter, supported by a variety of systems and batteries. They called it *Alvin*.

In 1966, when an air force B-52 bomber crashed and lost a hydrogen bomb in the ocean near Spain, *Alvin* proved itself by locating and lifting the device from nearly 3,000 feet. Before it reached the surface, however, *Alvin* dropped the bomb, which fell back down the seafloor to an even deeper, unknown position. The navy called on an experimental remote vehicle called CURV (Cable-controlled Undersea Recovery Vehicle) to do the final recovery. None of it was clean or neat, but it further highlighted the need for technologies to enable access to deep water. As at Skerki Bank, human and remote systems operated side by side.

Alvin was part of a broad landscape of Apollo-era projects that brought human presence to the seafloor, much as Americans were projecting that presence out into space. But while the Apollo program was receiving billions of dollars to send men to the moon, *Alvin* was struggling to raise a few million to bring humans to an equally unexplored world.

Ballard, as the point person between the navy and Woods Hole, became intellectually engaged. In the post-Vietnam drawdown of the armed forces, he left the navy and joined the *Alvin* group at Woods Hole, tasked with finding new customers for *Alvin*—what he called being a "scientific research salesman." Ballard also began attending the University of Rhode Island to earn a PhD in marine geology; when he finished his dissertation, he joined Woods Hole as a research scientist.

During the late 1960s and 1970s, *Alvin* began conducting scientific expeditions, gradually venturing farther from home, becoming more reliable and robust, more maneuverable, and better able to carry proper scientific instruments, samplers, and manipulation tools. Its depth rating was also increased, to 4,000 meters (13,124 feet).

Alvin's engineering history paralleled the emergence of plate tectonics, the formal scientific theory that emerged in the 1960s from ideas about continental drift that had been developing throughout the twentieth century. The young field of marine geology was at the forefront, providing magnetic, bathymetric, and seismic data to support the idea that the seafloor might actually be the place where the earth's crust was forming. The crust was being created along the mid-ocean ridges, as the plates were pulled apart by the ocean's trenches, where the crust falls back into the earth's interior.

But scientists had little direct evidence from the environment to support the idea of seafloor spreading. Traditional practice in oceanography, which involved dangling instruments from a ship or dredging, lacked the precision required to characterize and sample the mid-ocean ridges.

A series of joint U.S. and French dives in 1973–74, dubbed Project FAMOUS, brought two French vehicles and *Alvin* to the mid-ocean ridges to map and collect samples. The project provided crucial confirming evidence of plate tectonics. A rich era of exploration was born.

Moreover, Project FAMOUS brought *Alvin* away from shore and into deep water for the first time, where it proved its worth as a scientific tool. "It was fortunate that *Alvin* was there," Ballard recalled, "at the time that plate tectonics really took off." Nonetheless, even after FAMOUS, Ballard felt that much of the scientific community still regarded *Alvin* as a "gimmick."

Key to establishing the scientific credibility of *Alvin* was combining its use with another technology—a network of acoustic transponders placed by the crew around the area of study, which could navigate the submersible on precise x,y coordinates. Recently developed at Woods Hole under an Advanced Research Projects Agency (ARPA) contract, the transponders were battery-powered instruments deployed from the ship at the start of a dive series. They each listened for a ping, then returned a ping of another frequency after a fixed time delay. By "interrogating" these transponders and then listening for the replies, *Alvin* or the ship (or any other device), could fix its position within the field. The navigation data gave *Alvin*'s explorations a quantitative baseline, allowing scientific samples or observations to be located precisely within the ridge system.

Ballard developed methods to bring *Alvin* data to bear on scientific hypotheses. He saw *Alvin* as allowing scientists to replicate the practices of field geology deep in the ocean. "The key observational factors involved are the scientist's trained mind, his eyes, and the hammer in his hand," he wrote. "And the nimble little white submarine allowed WHOI's marine geologists the vital element of observational presence on the ocean bottom."

Even with the additional data from Project FAMOUS, the heat flows of the ridge systems didn't all add up, and scientists began postulating that enormous amounts of heat must be vented from the earth somewhere up onto the seafloor. A likely mechanism was that seawater pressed into the crust by the overlying pressure heated, then ejected back up. On a project in the Galápagos Islands in 1977, Ballard and a team of scientists documented the presence of exactly these hydrothermal vents near the islands.

They found not only the vents, but also unusual ecosystems around them, teeming with life in the deep-ocean environment, which previously was thought to be biologically desolate. The trouble was, the expedition had been designed to investigate the geology of the ocean floor, and despite the astonishing discovery, there were no biologists on board.

Ballard arranged hasty, long-distance single-sideband radio conferences with biologists Holger Jannasch and Fred Grassle back in Woods Hole. "We asked Holger and Fred how to proceed," Ballard recalled, "in effect trying to compress four years of undergraduate biology fieldwork into one scratchy radiotelephone call." Even this crude communications network could bring new expertise to bear on the seafloor.

Project FAMOUS and the Galápagos expeditions used another, less glamorous device alongside *Alvin*, one that sowed the seeds for the robotic future. Before *Alvin* dived, a camera sled dubbed ANGUS (for Acoustically Navigated Geologic Underwater Survey System) was towed behind the ship with a long cable.

ANGUS was merely a rugged steel frame about the size of a car, enclosing a set of 35mm color still cameras and flash strobes set to fire every ten seconds or so. While towing, the surface crew had only the most basic feedback (from an acoustic altimeter) of the sled's height above the bottom, and they adjusted the cable payout to keep ANGUS about 4 meters (13 feet) up, the ideal altitude for photographs.

The terrain on the mid-ocean ridges could be pretty hairy, often rising up faster than the humans could adjust. ANGUS's heavy frame was designed to bash into things and survive with minimal damage to the cameras. After several hours of survey and abuse, ANGUS would be brought on board and its color film processed. The ANGUS team's motto was "Takes a lickin' and keeps on tickin'," echoing a popular advertising slogan of the day for Timex watches and likening the sled to a clockwork automaton. They lovingly dubbed the vehicle "dope on a rope" because of its absence of sophisticated controls.

Yet towing a heavy sled behind a mile and a half of cable required a great deal of finesse. The cable, about the diameter of your thumb, seemed thin enough when descending in the giant ocean, but add up several miles of it and the surface area amounted to dragging the side of a building through the water. If towing were too fast, this drag would cause the vehicle to kite, flying too high up off the bottom to get useful data. Only achingly slow towing speeds and heavy weights could mitigate the kiting. Moreover, because of the drag, the motion of the ship took a long time, sometimes hours, to make its way down to the vehicle.

The team tried to tow the vehicle across the ridge in a series of straight, equally spaced track lines (often called "mowing the lawn," as we would do years later with NR-1). But running a straight track line when it took two hours for a correction to reach the bottom posed a frustrating challenge; turning the ship and the vehicle around at the end of a track line could take all day.

Most large ships simply could not move slowly enough, even running straight. But the Woods Hole Research Vessel *Knorr* had a unique pair of cyclodial propellers that allowed it to hover over one spot in the ocean, or move extremely slowly in any direction (later, these would be replaced by dynamic positioning systems, common in the oil industry, to accomplish the same function).

At first, both the winch and the ship were controlled from the *Knorr's* bridge, these being traditional maritime functions. The ANGUS crew, working from their control van out on a back deck, would call up "up one" or "down one" to operate the winch. These microinstructions eventually became tedious to the ship's crew, so they added a remote-control box for the winch right into the control van, and eventually a computer connection to the ship's control system as well. Even before robotics, these technologies called for adjustments in the placement and nature of human controls.

"We found running the sled [for] an hour was about all you could do," ANGUS group member John Porteous recalled, "and then you started getting a little goofy." The operators focused on a paper recorder that indicated the height of the sled above the bottom, and tried to operate the winch to keep it at the right height, which amounted to just a quarter inch on the paper. They also coordinated with the bridge to give the ship instructions on how to slowly proceed.

These early projects set the stage for later events and conflicts in some crucial ways. The remote system ANGUS operated in concert with the human system *Alvin*, often diving at night while *Alvin* was charging its batteries. The combination of acoustic navigation and precise positioning of the ship tied the whole thing together into an integrated system that could gather both qualitative and quantitative data from the seafloor.

Radio calls back to Woods Hole highlighted the contingent nature of exploration and expertise: if you are truly exploring, and don't know what you'll find, you may not have selected the right people before leaving home. A simple radio link could draw in a broader scientific community.

Finally, the "dumb" nature of ANGUS had limitations too—operators could not understand whether they had the right data, or any data at all, until they processed the film at the end of the dive. Sometimes entire daylong dives were lost because the cameras broke in the first five minutes, or were set for wrong exposures.

In later interviews, none of the ANGUS operators describe what they were seeing as "presence" on the seafloor. After the second Galápagos expedition, the ANGUS crews recalled their excitement, not at their own pictures, but at seeing the video *Alvin* brought back from a dive. In team member Steve Gegg's words:

> We were speechless! Even though we'd seen pictures of them [the deep sea creatures]. . . . But when you actually saw what people in *Alvin* were seeing . . . this was footage of what was actually going on down there. And it was just amazing. You know, the tube worm with all the shimmering water, and a crab walking up it . . . When you finally saw the video and you could understand what everybody was raving about, that made a lot more sense to me.

Twenty-five years later, what sticks out in Gegg's mind is not the *accounts* of the *Alvin* crew members, not their relating of personal experiences on the seafloor. Rather, it's the video, the imagery of things on the seafloor in motion.

ANGUS's limitations were not about *experience* but about *time*—the feedback from processing color film just took too long. If those images could come up quickly enough to change the next dive plan, then you'd really have something.

Ballard noticed a similar phenomenon—but actually inside *Alvin*. For the first two dives, the biologists and the *Alvin* pilots were not able to locate the vent sites. Even though Ballard was head of the ANGUS team for this expedition and not officially working with *Alvin*, the science team asked him to come over to their ship and dive in the sub to help find the vents. Once on the bottom, he found a fissure, started to see crabs, and "followed the crab gradient" until they came upon a vent field.

In his excitement, Ballard looked over at Holger Jannasch, the senior Woods Hole scientist who was the first biologist to see the world-changing discovery. "So we got there, and I will never forget," Ballard recalled of the scene inside *Alvin*, "and Holger has his back to the porthole, looking at the TV monitor. And I said, 'Holger, what are you doing?'

'I'm looking.'

'Why aren't you looking out there?'

'It's better, [the view is] better here. I can see better.'"

The color TV image from the RCA camera was so good that, even though he was physically present on the seafloor, the scientist was experiencing the seafloor remotely through a camera. "A light bulb went off in my head," Ballard recalled. "What are we doing down here?"

After that expedition, in 1980 Ballard took a sabbatical at Stanford University to write up his results for publication and prepare for his tenure review at Woods Hole. There, in the environment of a world-class engineering school, Silicon Valley, and the beginnings of the personal computer revolution, his mind turned toward the next way to explore the deep ocean.

"After using ANGUS to help map critical segments of the mid-ocean ridge," Ballard wrote, "I realized that a more advanced and sophisticated form of remotely operated unmanned vehicle could ultimately become a much more important scientific and exploratory tool than *Alvin* ever could be." His vision was one of replacement and competition—manned submersibles had a "questionable future" while remote vehicles would become "much more important."

A 1981 article in *National Geographic*, "New World of the Ocean," by Ballard's friend Samuel Matthews, surveyed ocean vehicles used for exploration. A full-page graphic portrayed various means for human presence in the deep ocean—most of them submarines and submersibles: the NR-1, the bathyscaphe *Trieste*, even William Beebe's original bathysphere, and *Alvin*. Remote systems included Scripps Oceanographic Institution's *Deep Tow* (which had contributed to the Galápagos expeditions), and a new, untethered French vehicle, *Epaulard*.

The graphic also included Ballard's new idea for a "two-body" system combining an ANGUS-like sled and a mobile, robotic vehicle. He called the system Argo/Jason after the mythical explorer and his vessel. A separate graphic focused exclusively on this vision, describing Argo/Jason as "robots that will carry not men but, instead, their eyes and ears as well as other sensors into the depths." Quoting an unnamed source (likely Ballard), the article says that "they will be more efficient . . . than such Model T craft as *Alvin* . . . [enabling] the scientist to project his eyes—his mind—into the abyss with perfect safety and with virtually no time limit on his 'dives.'"

A beautifully drawn graphic depicted the system: a ship, scanning with a bathymetric sonar ahead of sled *Argo*, a satellite link to shore (replacing the scratchy single-sideband link), *Argo* scanning with sonars and video cameras, and, emerging out of its "garage," a spherical robot *Jason* with two anthropomorphic hands gathering biological samples from a mid-ocean ridge.

On returning to Woods Hole (and receiving tenure) Ballard set about convincing the Office of Naval Research to support his vision, using his skills as a scientific supersalesman. ONR had long been a supporter of basic ocean science (hence Ballard's original job at Woods Hole), but it also had an interest in the ability to penetrate shipwrecks

Robert Ballard's early conception of telerobotic presence in the deep ocean, 1981. *Argo*, the sled suspended beneath an oceanographic vessel, scans the mid-ocean ridge while the remote robot *Jason* performs close-in inspection work. Ballard used this image to rally support and recruit engineers to his vision of remote presence in the deep ocean.

(COURTESY NATIONAL GEOGRAPHIC SOCIETY)

to see (or remove) what might be inside. Two nuclear-powered submarines, the *Thresher* and the *Scorpion*, had sunk in 1963 and 1968, respectively, and the navy wanted to know if they were leaking radiation. They also wanted to make complete maps of the wreck sites, something an *Alvin*-like vehicle would find difficult to do. Though not stated, clearly the navy would have an interest in surveying shipwrecks from other people's navies as well.

Ballard soon succeeded, receiving a commitment from ONR for $600,000 per year in return for a promise to let the navy use the system he would build one month per year (much of this money was actually provided in secret by naval intelligence, using ONR as a scientifically acceptable conduit). Ballard split away from the *Alvin* group and founded his own group, the Deep Submergence Laboratory, in 1982. The relationship between human and remote presence on the seafloor was not only between two separate technologies, but now two separate groups of people as well; they would not always get along.

In full-scale building mode for the new lab, Ballard began collecting equipment and people. His partner in the lab was Skip Marquet, a WHOI engineer who had long worked with the *Alvin* group and had developed its crucial acoustic navigation system. Now that *Alvin* was considered mature, Marquet was looking for new work. Ballard and Marquet brought the ANGUS veterans Tom Crook, Earl Young, Steve Gegg, and Cathy Offinger, who would be Ballard's right-hand person for years. They also brought in Stu Harris, an electrical engineer from Lockheed who had experience managing large projects in digital imaging for satellites.

The team would follow a staged process: first, develop *Argo*, with a coaxial cable, and then upgrade the system to fiber optics. Then build a small version of *Jason* to penetrate shipwrecks, then the full-scale version to integrate with *Argo*.

While Ballard was prospecting ONR for support, the agency suggested he make an alliance with MIT, where ONR was already supporting a researcher in telerobotics. Tom Sheridan was an MIT professor with an unusual pedigree. He had been a student of the behavioral psychologist B. F. Skinner as well as a mechanical engineer, and had long been interested in human interactions with machinery (Skinner's "black box" approach to behavior had always been appealing to engineers). The investigation into the 1979 nuclear accident at Three Mile Island had called on Sheridan's skills, and helped shape the nascent field of human-factors engineering.

Sheridan (who later became one of my thesis advisers at MIT) had been studying the interaction between humans and machines in real systems, and he realized that very few systems were truly manual, where the human controls everything by hand, or truly automatic, where the computer does everything. Rather, most systems existed somewhere in between, on a "spectrum of automation," and often shifted across that spectrum in real time. He termed this shift "supervisory control," where human and machine work together, trading control and shifting "levels of automation" to suit the situation at hand. Moreover, Sheridan recognized that precisely how supervisory control systems worked depended heavily on the social context in which they were embedded.

Ballard visited Sheridan at MIT, and met there a young engineer named Dana Yoerger. Yoerger had grown up in a family of shipyard workers, so when he went to MIT to study mechanical engineering he retained an interest in the ocean. Yoerger was attracted to Sheridan because of his broad purview. "He was very much interested in social issues outside his work . . . in the social side of technology. In fact, he didn't think you could understand technology without it." Yoerger had finished his PhD and was doing a postdoc in Sheridan's lab while he looked for a job.

Then Bob Ballard showed up. "Bob brought his stump speech" about Argo/Jason, Yoerger recalled. He showed Sheridan and Yoerger the Argo/Jason drawing from *National Geographic*. Yoerger remembers the moment of his conversion, as he looked up at the clock. "The meeting was at ten o'clock. It was [now] twenty after ten and I thought, this is what I want to do."

Yoerger signed on to the new lab, which Ballard was calling DSEL, for Deep Submergence Engineering Lab. When technician Earl Young started calling it "diesel" Ballard changed the name, because in the navy "diesel" represented the old world of nonnuclear submarines. It became just DSL.

To populate the new lab, Ballard needed more than fresh PhDs. Up the road from Woods Hole was a small company named Benthos, which had been started by Sam Raymond, a student of Harold "Doc" Edgerton at MIT. Edgerton had become famous for his electronic strobes that produced iconic pictures of bullets going through apples. He had also developed cameras and strobes for the ocean, which were marketed by Benthos.

Benthos had also been experimenting with simple, shallow-water tethered robots, without much commercial success. They donated one of their prototypes, known as RPV (for "remotely piloted vehicle") to Woods Hole.

Soon thereafter, WHOI was approached by the fire department in Quincy, Massachusetts, to find a missing teenager who was feared drowned in an abandoned, water-filled rock quarry. Ballard saw an opportunity to try out his new robot, and sent his two technicians, Tom Crook and Earl Young, to Quincy. They had years of experience going to sea and working in demanding deep-ocean environments, but the robot was new to them, so they also brought a young pilot from Benthos, Martin Bowen.

More than any other person, Martin Bowen created and exemplified the role of the remote-robot pilot in the deep ocean. Five years before he would fly a remote robot down the grand staircase of the *Titanic*, Bowen was a technician with a background in diving, biology, and photography. At Benthos he assembled production instruments, but found the new robots fascinating.

Benthos had a test tank the size of a large indoor pool, and Bowen spent evenings there learning to pilot the robot with precision. He darkened the windows of the tank, set up temporary scaffolding made out of aluminum tubes, and practiced flying in and out of these mock underwater structures using only the images from the vehicle's forward-looking video camera. The experience was novel because the field of view was so narrow.

Bowen made the analogy between this state of mind and his experience as a commercial diver. "I had to be aware that something was going to be nibbling at my swim fins. There's always something behind you, and that translated very easily into the 3-D world of underwater robots; now something's nibbling at my tether." For the unusual tethered motion, Bowen built up "an awareness of how you venture out, how you observe, record, and then retreat." In those quiet, solitary evenings in the Benthos tank, "I would play this Hansel and Gretel game where I would fly the vehicle in and out of the scaffolding and then try to work my way back so I wouldn't be tangled." As a photographer, Bowen also developed a sense for 3-D movement underwater.

Bowen, Crook, and Young went to the quarry and began searching. It was an eerie task; mostly, the bottom of the four-hundred-foot-deep quarry held old cars and shopping carts. But the prospect of the grim target loomed. "Every time I turned around with a video camera in the vehicle," Bowen recalled, "I thought I was going to see a pair of sneakers and a pair of shorts and a young face." They never found a body. Years later the teenager was found, alive and well, living in Texas. The disappearance had been a hoax.

But the Woods Hole team gained valuable experience; not only had they searched, but they also made a map, plotting their way around the quarry with ropes to estimate position and be sure they searched the full area. Ballard was impressed, and offered Bowen a job. Bowen was excited by the prospect. "These were doers," he said of Crook and Young, "these weren't people who sat in front of computers. These were people who went to sea. They made things happen." Ironically, they would become people who went to sea *and* sat in front of computers.

When Bowen arrived, Ballard's new lab was already cooking. *Argo* was well underway. Like ANGUS, *Argo* was a towed sled; Ballard called it "an improbable kite of white steel tubing the size of a station wagon, mounting a single jaunty tail fin." Unlike its predecessor, it sent telemetry data and real-time video straight up the cable, though like ANGUS it also took still pictures with film cameras. Fiber-optic cables with enough strength for oceanographic applications would not be ready for several years, so *Argo* used the traditional means of coaxial cable, similar to what you plug into the back of your cable-TV box. Electrical power, data, and video were all multiplexed on the same electrical conductor, a tricky business. Unless perfectly tuned and shielded, the signals could interfere with one another, causing noise in the data and snow on the video.

Dana Yoerger, the sole PhD engineer in the group, worked on understanding the dynamics of the long towing cables, and on improving the automatic, "dynamic positioning" of the surface ship. He also began setting up a research program in how to use automation to improve the vehicles down on the seafloor. If the ship could hold station with its position feedback, why not dynamically position the robot too? What new technologies would be required to make that work? How could such a supervisory control system help the pilot and the scientists?

By the summer of 1984, *Argo* was ready, and did a thorough survey of the site of the wrecked submarine *Thresher* in 5,800 feet of water. The real-time video feed from the sled was invaluable in locating the wreck and navigating across it.

Most important, the video finally gave the team the sense of presence Ballard was looking for. "Certainly all of us in the control van felt we were down there with *Argo*," Ballard wrote in his memoir. "Our scanning human eyes and restlessly curious brains had been transported to the ocean bottom. The rest of our vulnerable human bodies remained above the crushing depths in the comfortable air conditioned control van. . . . The screens seemed more like port-holes than television monitors." Sonar imagery and data from other sensors only enhanced the crew's experience.

The *Thresher* survey gave Ballard and DSL credibility with the navy sponsors. The following summer, ONR approved three weeks of *Argo* time to survey the site of the *Scorpion* as well, with the understanding that if the survey were completed early, the team could use the remaining time for other oceanographic projects. The *Scorpion* survey was successful too.

From there, *Knorr* and *Argo*, again with a collaborating French expedition, sailed north for these "other projects." For Ballard that meant pursuing an old dream of searching for the *Titanic*.

By August 31, 1985, after a week of painstaking searching from thirty feet above the north Atlantic seafloor, twisted metal wreckage, and then a recognizable boiler, appeared in *Argo*'s fuzzy black-and-white video in the control van. The lost wreck had been found.

In the exciting days that followed, *Argo* would fully document the

Titanic wreck with photographs, video, and measurements. ANGUS too would get into the act, ticking through thousands of 35mm color stills. The resulting mosaic, hand-assembled with scissors and glue by Ballard's mentor, geologist Al Uchuppi, would be published in *National Geographic*.

The discovery of the *Titanic* wreckage would become known as the greatest undersea find of the twentieth century, propelling Ballard, undersea vehicles, and Woods Hole to new heights of fame. It also opened the deep ocean for an eager public imagination. Since ancient times, the deep ocean had been black and unknowable; remote vehicles now rendered it visible and accessible to humans.

When the *Knorr* returned home after the *Titanic* discovery, thousands of families and well-wishers crowded the docks and the tiny town of Woods Hole. After Ballard and his crew disembarked from the ship, WHOI held a press conference at a nearby auditorium, packed with reporters from all over the world.

Here, at his moment of greatest triumph, a moment that would transform his life and career, Ballard used the opportunity to press his vision of remote presence. He maintained that *Argo* and its cousins under development at Woods Hole represented "a complete revolution" in underwater exploration. "It's the beginning of telepresence, of being able to project your spirit to the bottom, your eyes, your mind, and being able to leave your body behind. . . . We've entered a new era in undersea exploration."

That *Titanic* was discovered by remote presence, and not by physical human bodies on the seafloor, however, would long be a source of tension within the Woods Hole Oceanographic Institution.

That tension would become a literal tug across the seafloor when Woods Hole returned to the *Titanic* the following year, this time to enter the wreck and explore inside. DSL received from the navy a small vehicle with the awkward name AMUVS (Advanced Maneuverable Underwater Vehicle System), a round robot designed to fit into a submarine's torpedo tube and be jettisoned out for odd jobs.

AMUVS was called a "roving underwater eyeball." Indeed, that phrase came to stand for the most basic use of an ROV—to move an underwater video camera around at the will of a human pilot from some other location. It had been mounted on the exterior of the bathyscaphe *Trieste*, but the installation was clumsy and didn't work very well. AMUVS did spawn a commercial product, RCV-225 (remotely controlled vehicle), an early ROV.

AMUVS was a mediocre design, but it had one extraordinary feature—it was built inside a beautiful, oddly shaped pressure housing made out of a single piece of machined titanium. About the size of a desktop computer, it housed all the electronics, capped with a quartz crystal hemisphere to accommodate a camera lens. This pressure housing would become the core of a new vehicle.

Ballard's DSL engineers ripped out AMUVS's guts, put in a new video camera that could pan and tilt using model aircraft servos, built it into a new piece of underwater foam for flotation, and added the newest electric motor thrusters. This vehicle became known as *Jason Jr.* (Ballard already had plans for the full-size *Jason*), and its goal was to demonstrate that a remote vehicle could penetrate the hull of a sunken ship and find something interesting inside.

Ballard's story about remote vehicles making manned submersibles obsolete did not go unnoticed within the *Alvin* group. On one level, Ballard was simply in his salesman mode, preaching a narrative of progress that was deeply ingrained both in his navy sponsors and in public discourse. On another level, he was playing bureaucratic politics—the internal lab versus lab rivalries at which academics excel.

In an interview with the *Cape Cod Times*—Woods Hole's local newspaper, of course—Ballard predicted, "Manned submersibles are doomed." Ballard later called the comment "regrettable," but still felt that after *Argo*'s discovery of *Titanic*, "the *Alvin* group lost some of its popular glamour."

He soon had a chance to eat his words.

Ballard's lab was now funded by the navy to send *Jason Jr.* inside the sunken wreck of the *Scorpion* to find its nuclear weapons. In 1986, as a cover story for that secret project, Ballard engaged *Alvin* for a more detailed exploration of the *Titanic* wreck, including use of the new robot *Jason Jr. Alvin* was to ferry the steamer-trunk-sized *Jason Jr.* down to the wreck in a garage mounted on the sub's front end, whence it would venture inside the ship. A fiber-optic cable connected the robot to a remote pilot inside *Alvin*, who would also operate the camera and record video. After the wreck's initial discovery via remote vehicle, the *Alvin* group was pleased to be brought in for the serious investigation.

Once the upstart, *Alvin* was becoming the establishment. In 1973, a formal *Alvin* review committee had set up procedures for peer reviewing proposals and allocating time on the submarine—providing formal academic process. In 1984, the improvised support platform *Lulu* was replaced with a proper oceanographic vessel, the *Atlantis II*, as *Alvin*'s mother ship. By 1986 *Alvin* had logged more than 1,700 dives and was proving itself as a reliable scientific tool. It would dive more than a hundred days in 1986 alone for a broad variety of science users. Its pressure sphere was now made out of titanium and rated to 4,000 meters (13,000 feet).

Also in 1986 a full electronics overhaul replaced the hydraulic motors with brushless DC motors for propulsion and control. The pilots finally got a proper joystick to fly with, rather than having to flip toggle switches for individual motors. "*Alvin* became a different machine with that transformation," pilot Will Sellars recalled. His first experience with the new setup would be on the *Titanic* wreck.

By contrast, *Jason Jr.* now played the cutesy upstart. Only just built and barely tested, it had never been tried in deep water, and had been mounted on *Alvin* a mere three days before the ship left port in Woods Hole. As Martin Bowen put it, "We went from [testing in] ten feet of water to two and a half miles."

Ballard felt the *Alvin* group saw him as returning with his tail between his legs after publicly dissing the submersible. Evidently, some in the *Alvin* group felt that way too: on the way to the dive site, the cook on *Atlantis II* made a special cake for Ballard, decorated with his words "Manned Submersibles are Doomed." In a light moment that barely masked the tension, they made Ballard literally eat his words.

Despite the humbling, the dive series would do nothing to resolve strains between the proponents of manned and unmanned systems; indeed, it would highlight them.

In Ballard's eyes, the *Alvin* pilots were skeptical of the ROV. During the dives, he felt *Alvin* pilot Ralph Hollis was taking risks with the sub just to show it could equal *Jason Jr.* in its ability to probe the wreck and return data. Yet Ballard found *Alvin* procedures hidebound. The sub had to leave the bottom around three in the afternoon every day to return to the surface so the crew could be on board by dinner time. This schedule was dictated by safety concerns and the need for the crew's rest and overnight maintenance time, but Ballard found it "bureaucratic inflexibility that was undoubtedly exacerbated by a good admixture of professional jealousy." In Ballard's mind, "*JJ* was in the process of proving the superiority of ROVs over manned submersibles."

In interviews, the *Alvin* crews didn't mention this comparison. Secure in their newly established status, they viewed the ROV as an annoying, inferior toy.

They were right. On the first dive, "When the *Alvin* pilot turned on the science [electrical] bus to power up the vehicle," Bowen recalled, "this thing [*Jason Jr.*] started to cook, and it looked like a little white plume coming out the back of the vehicle." Electrical junction boxes shorted and smoked out. Thrusters failed. On one dive, when recovering *Alvin* on the surface, little *Jason Jr.* fell out of its garage, a true "dope on a rope" swinging below *Alvin*. *Jason Jr.* had to be rescued by divers, who cut the tether. Of the twelve dives, *Jason Jr.* got good video on only a fraction of them.

Alvin's first descent to 14,000 feet took about two and a half hours. Ballard and Bowen sat cross-legged inside, their laps full of boxes of videotapes to record *Jason Jr.*'s visual treasures. The sub would land near the wreck and then gradually inch its way in until it saw the huge target on a sonar screen. Then, approaching within visual range, "in the portholes you'd see an eight-story building, and you're at parking-meter level. So you see this thing rising above you."

Ballard's dream was to fly *Jason Jr.* down the grand staircase of the *Titanic* and record the adventure on video. This maneuver required a delicate dance between *Alvin* and the robot. The submersible had to land on the deck, ballasting itself slightly heavy to create a stable platform. Would the decks hold? Or would they give way, potentially swallowing both vehicles? Pilot Ralph Hollis tried landing as a test, and the decks seemed to be sturdy enough.

On the third dive, pilot Dudley Foster, a mechanical engineer and former naval aviator (who would go on to become *Alvin*'s longest-serving pilot), laid the sub on the deck and ballasted about twenty pounds heavy. Once *Alvin* was in place, Foster had to inch the sub as close as possible to the giant staircase hole, because *Jason Jr.*'s reel had only about 150 feet of tether cable. *Alvin*'s viewports were angled down at about 45 degrees, so the pilot couldn't see much directly in front of the sub. When Foster got close enough to the hole, "all I could see was that big gaping hole out the front window."

Bowen had a control box in his lap; a finger joystick on the right let him drive the vehicle horizontally, and a toggle switch on the left thrusted up or down. In the middle, a prototype Sony Watchman portable TV showed a black-and-white video image from *Jason Jr.*'s nose camera.

Inside the sphere, Ballard sat nervously while Bowen stared intently at his little Sony Watchman. Bowen eased *Jason Jr.* out of the garage, slowly feeling his way forward. If he moved too quickly or brushed the wreck, he could kick up a cloud of dust and lose his orientation.

All the participants were aware of the razor-thin line between *Jason Jr.* as an advanced robot and *Jason Jr.* as an expensive, deadly anchor for *Alvin* and its crew. The tether was rigged with a cable cutter to allow the sub's escape in case of emergency, but no one was sure whether it would work. "If I penetrate the wreck with this strong tether," Bowen recalled, "and the vehicle died or something failed catastrophically, we'd be there"—possibly for good.

At one point, the bright yellow tether, about a half inch in diameter, did catch on a piece of broken railing. Bowen didn't realize there was a snag, and he flipped the switch to reel in the cable. Instead of reeling *Jason Jr.* back toward the sub, the stuck cable pulled the sub toward the dark hole. The submersible and the ROV formed a coupled dynamic system, and a coupled human and remote system. Now, through this taut cable, the tension between manned and unmanned became literal. Realizing his error, Bowen switched off the winch, flew the remote vehicle back to pull gently on the tether, side-slipped a little, and pulled it out of the railing.

Despite the glitches, as *Jason Jr.* descended vertically into the staircase hole, Bowen lost himself in the robot. "I was in that vehicle's eye. I was looking at columns down in the staircase. I was trying to back up, move forward. . . . It was kind of a game." All the while, the *Alvin* pilot monitored the vehicle's systems and sensors.

Bowen had to simultaneously project himself through the front of the remote vehicle while also managing the demands of the human system around him. "The pilot would say, 'Get that thing back here!' or 'We're done, we're out of there. Turn the winch on.' Or Bob would say, 'OK, things are working. Go over there. Go over there. Inspect that.' And so I would try to balance all those demands."

But still he found himself transported. "I was just flying this thing. My brain wasn't inside the submersible. It was in the camera," as he flew inside this massive shipwreck nearly three miles down. Twenty years later, air force operators will say the same thing about operating drones in Afghanistan, and geologists will say the same thing about operating remote vehicles on Mars.

At one point Bowen pivoted *Jason Jr.* to look back at *Alvin.* "As we sat inside *Alvin*," he recalled, "suddenly headlights go by, like a dark night in the summer when a car drives by your bedroom, headlights go by the inside of a dark sphere. And once again peoples' legs sort of shot out and—What was that? Well, that's us. Oh, OK, that's us. Just trying to get a handle on who we were at that point." On another dive, pilot

Will Sellars was amazed at how just having *Jason Jr.* as a remote set of lights enabled him to see much farther out the porthole than he could have otherwise.

The dive down the staircase had lasted little more than twenty minutes when Foster pulled the plug and ordered the ROV in and *Alvin* back to the surface for dinner. But the video was in the bag—the key twenty minutes of the entire expedition.

The *Alvin/Jason Jr.* combination was the feature of a new *National Geographic* special program. Moreover, the *Alvin/Jason Jr.* combination

Tension between human and remote presence on the *Titanic* expeditions finds its way to the national press, 1986. This image appeared on the cover of *National Geographic*, depicting the small robot *Jason Jr.* peering into a window on the *Titanic*. By contrast, the cover of *Time* showed only the *Alvin* submersible, even though both vehicles explored the wreck together.

(COURTESY NATIONAL GEOGRAPHIC SOCIETY)

appeared on the cover of *Time* magazine, artfully portrayed by painter Ken Marschall. Curiously, only *Alvin* is visible on the *Time* cover, which is perhaps why many people think *Alvin* "discovered" the *Titanic*. By contrast, the *National Geographic* cover portrays only *Jason Jr.* peering into a window.

As with the discovery of the wreck a year earlier, the exploration inside the *Titanic* captured the public imagination regarding the deep ocean and the potential of robotics. A few years later, a dramatized version of the exploration of the hull became the opening scene for James Cameron's film *Titanic*, the second most popular film ever made. In that instance, the robot served not only to open a window into the shipwreck, but into the whole historical drama that led it there.

Jason Jr.'s sojourns around and inside *Titanic* represented the apotheosis of the early, "roving eyeball" phase of deep-ocean robotics. Those eyeballs could do a great deal, not least of which was to shift the risk— as *Jason Jr.* did for the people in *Alvin* when it ventured into a place too small and too dangerous for the full submersible. Yet these tasks were similar to *Alvin*'s early "go and have a look around" phase, before acoustic navigation and new scientific practices made it an established scientific tool, producing formal data sets.

The human and remote vehicle combination held no great attraction for the *Alvin* group: after *Titanic*, it was never used again.

Ballard and his lab still had some distance to go to fulfill the original Argo/Jason vision of telepresence on the seafloor. During the next few years, the full-size *Jason* took shape and underwent a series of shallow-water trials. *Argo* was rebuilt from a camera sled to a full-size garage, capable of fully enclosing the Volkswagon-sized *Jason* within itself. The new *Argo* garage was dubbed *Hugo*, for "huge *Argo*." The new vehicles had fiber-optic cables, adapted from telecommunications systems, capable of transmitting gigabytes per second up the long distance with no degradation in quality.

Yet it was not merely as roving eyeballs that remote robotics would have their impact in the deep ocean. Rather, these vehicles quickly became mobile digital sensor heads, capable of vacuuming enormous amounts of data from the seafloor and sending it to the surface. They would not be fully automatic, nor fully manual (no *Alvin* toggle switches here), but rather operated by the supervisory control systems Dana Yoerger and his team were developing for *Jason*. Nearly all the "smarts" would be located on the surface—*Jason* itself was little more than a platform, providing power, telemetry, and mobility to any number of cameras and sensors bolted onto its structure.

Jason first went to sea in September 1988 in the Hood Canal in Washington's Puget Sound, doing shallow-water survey work for the navy. The spring of 1989 saw the first deepwater test, on an archaeological project in the Mediterranean (the predecessor to the later NR-1 expedition that located Skerki D).

The thrill of seeing a decade's worth of work slowly lowering its way to the seafloor did not last long. Once in the water, the *Hugo/Jason* combination had a large mass but low weight. When the ship heaved up and down, the cable would go slack, then on the next crest violently snap tight again. On Ballard's proud new Hugo/Jason system's third dive, the cable snapped at the terminal connecting it to *Hugo* and the whole shebang fell quietly to the bottom in 800 meters (2,600 feet) of water. Perhaps this was the first inkling that ROVs might not necessarily be safer than the human-occupied submersibles.

In a series of heroic improvisations, the DSL team grappled *Hugo* and *Jason* from the bottom. *Hugo* was discarded for good, replaced with a miniature camera sled dubbed *Medea* (after the legendary Jason's murderous wife), that has survived to this day. No longer serving as *Jason*'s garage, *Medea* acted as a heavy weight to stabilize the heaving from the long cable to the surface. *Jason* connected to *Medea* through a neutrally buoyant tether only about 150 feet long and was free to move precisely within this radius without being affected by the ship's motions far above. *Medea* also had camera and navigation systems on it to keep a kind of bird's-eye view of *Jason* as it worked.

The impromptu modifications to the Argo/Jason system brought home an unexpected difference from the human systems. *Alvin* had to pass a rigorous series of tests from the navy to be certified safe to carry humans. These included everything from tracking materials that made up the hull all the way back to their sources to approvals for any engineering changes, which could only be made at biannual overhauls. To this day, there is no software in the life-critical systems of *Alvin*—it's an old-fashioned piece of hardware, simple and safe.

By contrast, remote vehicles were not subject to certification requirements, so they could be modified easily, cheaply, and in the field. Moreover, because the software did not need to be certified, it could be modified continuously during operations. *Jason* went through numerous hardware modifications, and constant software revisions, during its first few years. Software crashes were a common occurrence in *Jason*'s early days, some of which would have imperiled a human crew, were one on board.

On a summer day in 1988, two years after the *Titanic* exploration, I walked down the stairs of an old, green aluminum building in Woods Hole with a small painted sign out front: "Deep Submergence Laboratory." Looking for a job, I was there to meet Skip Marquet, one of the original *Alvin* engineers and cofounder of the lab with Ballard. Touring the lab, I saw exotic robots, heavy-pressure housings, and other things foreign to me. "This has been inside the *Titanic*," Marquet said as he pointed out *Jason Jr.*, opened up on a lab bench with its electronic guts pulled out.

But inside those robots, and surrounding them, were things deeply familiar—electronics, microprocessors, software manuals. In a moment, I was hooked—I could bring my skills and passions to this unusual, alien endeavor. Eager to travel the world doing engineering, build electronics that would find their way into extreme places, and not have to report to work in a cubicle, I joined the Deep Submergence Lab as a junior engineer.

With the basic *Jason* system proven, Yoerger could turn his full attention to developing more advanced supervisory control technologies to aid remote explorations. He hired me, an electrical engineer, to help develop the electronics and embedded systems for that automation. Specifically, I was to work on two projects: a precision navigation system that would use ultrasonic pings to navigate *Jason*, and the computational guts for a new "autonomous" vehicle called ABE (although I would work on other aspects of the *Jason* system along the way). Over the next several years, we deployed *Jason* to a variety of projects, each time writing new code and doing new things with the control systems, and each time learning something about remote exploration.

What was it like to operate a robot in the deep ocean? First of all, we should qualify the term "robot." The term was commonly used for the vehicle, but there was very little resembling autonomy about the vehicle. In fact, it was something of a blank slate, technically speaking. There was relatively little computing power on board, only enough to flick the lights and instruments on and off, turn on the thrusters, and do a little other housekeeping. Video signals went straight up the optical fibers, and most of the instrument data was simply multiplexed through the computer on *Jason* to go up to the top for processing. Even when *Jason* was doing something "automatic"—such as holding at a constant depth—the feedback loops were closed through a computer on the ship.

Physically, too, it looked neither sleek nor human-like. The top half was a solid block of specialized foam for floatation while the bottom half was a mess of railing, pressure housings, and wires. Parts of it were held together by duct tape and hose clamps.

Despite the grandiose, human name, very little of the robot was anthropomorphic, but if you squinted you could begin to make out a kind of face at the front where the cameras were concentrated and the manipulator arms did their work. A better analogy for *Jason* was a telescope—it was something you looked *through*. In fact, when it worked best, it was invisible, as it were, allowing the people on the surface to see the seafloor and forget about the medium of transmission. This was the key to the sense of presence.

Jason also drew the surface ship—one of humankind's oldest mobile, manned vehicles—into a closer relationship with the robot. Once *Alvin* descended, the ship was free to do as its crew pleased, and could move off the site to collect some other kind of research data. With *Jason*, the ship and the robot were always tied together by the cable, and had to be precisely coordinated. We would wire a computer connection from the control van up to the ship's positioning system, so the navigator in the *Jason* van on the back deck would have control of the ship. One click on the PC keyboard, and the ship would move a few meters forward, or to the left, or in a slow straight track line.

Down below ran the big cable, then *Medea*, and then *Jason*. A lot to keep track of—displayed on a navigation screen running a custom piece of software Dana Yoerger wrote and then customized for each trip and modified constantly during a cruise. "It's a ballet," Martin Bowen observed; he likened it to an arcade video game, except instead of putting in quarters every few minutes you were putting in thousand-dollar bills.

Jason's control room consisted of five or six twenty-seven-inch video and computer monitors mounted on the wall, displaying imagery from *Jason*'s multiple cameras, and navigation data. A series of control stations were arranged beneath them—one for the pilot, one for an engineer who monitored the vehicle's systems and controlled the cable winch, and one for the navigator. A data logger changed videotapes and logged all activity. This left plenty of room in the van for a chief scientist, who usually sat directly behind the pilot to direct the dive, and ten or twenty other people—other scientists, engineers, graduate students, and film teams from the media.

When all was stable, though, the whole control van would become concentrated on the seafloor. "Now, that's the world of telepresence," Bowen said. "That's where I forget about my body, and I project myself onto the ocean floor, and I have to make that vehicle dance."

The pilots found the immersion similar to diving in *Alvin*, but the social space in which they operated was different. The crowds in the van could certainly be distracting. "You become overwhelmed with input," *Alvin* pilot turned *Jason* pilot Will Sellars said. "You already are gleaning a ton of information off of these screens. And this person here, and you're trying to deal with your two guys here." In Bowen's words, "With robots you could have a whole gallery of experts behind you . . . They have their own disciplines that they would like to have satisfied for your dive. And I have to digest those and decide where I'm sending the vehicle next."

The pilots learned to narrow their attention to hear just a few voices. If the compass didn't look right, or the surface weather was flaring up, the pilot could ask the navigator about it. I stood the navigation watch, and learned to anticipate much of what the pilot needed and when he needed it: move the ship a bit; inform on the navigation quality; watch out, you're getting a little close to this. Gradually, the pilot would build up a mental picture of the terrain. "I just start mapping things in my own head, what the obstacles are, how high I need to fly, how low . . . I have the advantage of being surrounded by people who are also helping take care." Martin Bowen and I became close friends through many such shared experiences at sea.

New fiber-optic cables, digital transmission, and the very best broadcast-quality video cameras helped transport us into this other world. But this presence, or what Ballard called telepresence, was not simply a product of the technical imaging. In fact, by today's standards the imaging was primitive, being old-fashioned color TV quality rather than digital high definition.

Presence on the seafloor deeply related to what was going on in that darkened control room. As *Jason* pinged and photographed its way around shipwreck sites, hydrothermal vents, or other sites, the group in the control room was in constant conversation, observing, questioning, speculating on what the cameras and sensors showed. This constant, real-time seminar about the ongoing exploration, combined with the beautiful, haunting images we were seeing, is what transported us into another world. This was the most fundamental, surprising difference from *Alvin*.

Sometimes, if someone would bump into your chair in the control room, for a moment you'd be convinced that the robot down below had crashed into a rock—until you "woke up" and untangled your body from your mind. This is what the manned submarines could never replicate;

this was what the robots did for us. While our bodies have never physically been to many of these sites, our minds and imaginations spent days
and days there, and we had a deep sense of their underwater landscapes.

The changes brought with them professional tensions. Archaeologists, for example, accustomed to digging or diving in their own holes,
had difficulty adapting to a world where they did not control the means
of access to their sites. It took some convincing from us for them to
realize that actually sitting in the pilot's seat and operating the joystick
for *Jason* would not be the best use of their talents.

Jason, of course, did no archaeology at all—it merely passed the data
it sensed up the cable. Once uploaded, engineers and programmers
assembled the maps and mosaics, but it still remained for the archaeological team to provide a sense of interpretation. They were not exploring the seafloor with their bodies. They were exploring, but they were
exploring in the data, sitting in front of a computer. Some archaeologists
found it thrilling, others repulsive.

The question of professional identity was not limited to archaeologists or scientists. On one trip, an identity issue involving photography
came to an absurd conclusion. While exploring the wreck of the *Lusita-
nia* off the coast of Ireland, we had a photographer on board from
National Geographic who had all the ego you might expect of a represen-
tative of such a high-profile magazine. He was snapping away on the
ship, capturing the surface operations, but he realized that the real
action was happening on the seafloor, where he was not likely to get a
photo (or a photo credit).

He first proposed actually driving the vehicle to set up the shots,
but that idea, naturally, was rejected by the pilot and management as
unsafe for someone with neither training nor technical background.
But *National Geographic* was one of the "customers" for the trip, having

put up a large amount of money to make the film and magazine story about it, so they had some sway.

Eventually we negotiated a solution. I removed the camera button from the pilots' console, soldered it onto a six-foot-long cord, and wrapped it in tape to make a little handheld trigger. Even though the ROV pilot loaded the film, set up the cameras, and actually maneuvered the vehicle to frame the pictures, the *National Geographic* photographer pushed the button to trigger the camera. Sure enough, when the story about the expedition came out in the magazine, the photographer, not Martin Bowen or the other ROV pilots, got the photo credit.

Over the next several years, our group continued to operate *Jason*, refining and developing it, and worked to gain its acceptance as a scientific tool. Ballard was in a unique position, as his relationship with the navy and *National Geographic* allowed us to undertake interesting, unusual, and innovative projects that would never have been funded through the traditional channels of scientific research, especially the conservative National Science Foundation. The Cold War had just ended, so the navy had some assets available for experimentation (like NR-1). We all knew that the situation was temporary; Ballard was willing to lend his support for a while, but he had his own agenda and would eventually move on to his next big thing. His relationship with WHOI was also continually strained, and he was in the process of establishing a new center in Mystic, Connecticut, to showcase his own projects. *Jason* would need to establish itself with mainstream oceanographers if it was to be viable in the long term.

Unexpected differences with *Alvin* emerged. The *Alvin* group was an operations organization—they operated an existing, certified vehicle. Every eighteen months or so *Alvin* would come in for overhaul, and upgrades were added, but often the philosophy was "don't touch what works." Gradually, over time, this allowed the sub to fall more and more behind current technology, especially in the arenas of computers and advanced sensors.

Jason, on the other hand, emerged from DSL, a research and development group. Because the sub was not certified by the navy, it could always be experimented with, always altered. It was in a constant state of flux, and usually we did not complete any one revision until the ship was underway to the dive site.

The scientists were hesitant. The technology was unproven. Some early trips were so plagued with technical difficulties that they returned only minimal data. Early expeditions even saw life-threatening disasters— one time a crane fell on us while recovering *Jason* in the middle of the night; another time the long cable (carrying deadly high voltage) unspooled on deck during rough weather in the north Pacific. Nothing about the new robots was proving either cheaper or safer than *Alvin*. Even the expeditions that went well had brief days of sparkling success offset by many days of problems and down time.

Most important, most scientists just didn't see how such a robot could accomplish their science. "People would say, ROVs? Pfff! What are they going to be? That'll never work!" Bowen recalled. They had grown fond of *Alvin*, with its enclosing structure, its cute name, and its decades of reliable operation (though some of the older scientists surely remembered the early decades, when *Alvin* was unreliable, idiosyncratic, and prone to loss). Remote science seemed to threaten the professional identity of explorer-scientists. Why would they willingly give up their exciting visits to the deep?

Another member of the team recalled the difficult, grinding work of "breaking down barriers in the eyes of prospective users who could just not see what was totally obvious to an engineer. I mean, why do you have to explain why an ROV would be useful in the deep ocean?" He thought it was "a social thing . . . a union kind of thing." The union motto? "Protect *Alvin*." It had been a twenty-year struggle to keep *Alvin* alive, and ROVs were seen as taking a piece of *Alvin*'s pie.

Will Sellars had an interesting perspective on this struggle. In 1989 he switched from being an *Alvin* pilot to being a *Jason* pilot. On an *Alvin* cruise lasting about three weeks, each pilot would dive about five times in a twenty-dive series, usually without the same scientists in the sphere more than once or twice. He felt that as a *Jason* pilot, he developed closer relationships with scientists "because you dealt with so many more of them."

Alvin was designed as a search tool (something it was never very good for). The views from the three viewports did not overlap at all, in order to give the three people's eyes maximum coverage of the seafloor at any given time. But most of the time *Alvin* did close-in sampling and manipulation work—and the scientists could not see what the pilot was doing except by taking a turn looking through his viewport, which interrupted the work. Or they could look at the video camera, as Holger Jannasch had done on the Galápagos expedition in 1979. "And you find people that go through a whole *Alvin* dive," Sellars recalled, "and experience the whole thing on that four-inch monitor right above their viewport." Remote presence proved compelling even within *Alvin*'s sphere.

Despite the rivalry, over the next several years, we continued to combine the remote and human vehicles in more experimental comparisons of the experience and the quality of the data. Off the coast of Seattle, at the Endeavor Ridge Hydrothermal Vents, I dived in *Alvin* to set up my navigational instruments around the vents. From the sub we saw stunning images of shimmering water floating upward against the bottom of the vents, water so shiny it looked like smooth, undulating pools of mercury, punctuated by brief black plumes that would pop down like inverted puffs of cigarette smoke.

Less than a week later, we returned to the site with *Jason*, which also took some great video. But our most proud technical accomplishment of the trip was using *Jason* to chemically explore the plume. We made *Jason* fly, via computer-controlled autopilot, through the hydrothermal plume (analogous to a smokestack) to make precise chemical measurements. This survey produced a three-dimensional chemical map of the plume, which enabled chemist Russell McDuff to measure, for the first time, exactly how the hydrothermal fluid mixed with the surrounding sea water.

This was the first inkling of the technologies that would come to fruition while mapping ancient shipwrecks at Skerki Bank several years later—the consolidation of Dana Yoerger's vision of supervisory control. The combination of navigation, sensing, imagery, and precise computer control made the robot into something entirely other than the so-called roving eyeball or the human-occupied submersible.

Jason was digitizing the seafloor, creating virtual, 3-D models that could only exist inside a computer. "Intelligence" or "autonomy" were the least of its features; all of the engineering went into ensuring it would go only and exactly where it was told to go and nowhere else. What took intelligence was the human exploration that, instead of occurring inside the sphere of *Alvin*, now took place aboard the ship in the *Jason* control room, and even more extensively inside the data sets weeks and months after the expedition.

Yet the debates raged on, because they mattered. What kind of human presence should be cultivated on the seafloor? The question affected government science policy and funding. Officially, Woods Hole took the institutional position that "both are valuable," given that WHOI had major investments, financial and cultural, in both *Alvin* and *Jason*. But oceanography is a small, resource-constrained community, and couldn't afford to double the resources simply to avoid resolving the debate.

Alvin was nearly forty years old. Though most of the sub had been replaced in numerous overhauls, the basic design remained from the 1960s, and the titanium sphere was nearing the end of its life due to metal fatigue. Should *Alvin* be replaced with a new, fully updated version? What would a new *Alvin* look like? What new missions could it accomplish?

An influential group of scientists argued that a new *Alvin* should be built to dive to 10,000 meters, more than doubling the existing 4,500 meters deep. Given *Alvin's* illustrious history with hydrothermal vents, they argued, there was good reason to go deeper, at least to have a look around. They invoked national pride, as the Japanese now had a submersible called *Shinkai*, rated for 6,500 meters, and it was embarrassing to some that *Alvin* was no longer the deepest-diving submersible.

Of course if you looked at how *Alvin* actually spent its time, nearly all of it was between 2,400 and 4,000 meters, where the scientists studied the hydrothermal vents. The additional costs of a 10,000-meter rating were astronomical, and the idea soon floundered as it became obvious that remote or autonomous vehicles could survey the areas at much lower cost and with no lives at risk.

A second argument for the new *Alvin* was the quality of the imagery. "We all know how much better we can see from *Alvin* than from an ROV," the scientists argued. At that time *Jason II*, a larger and improved successor, was just coming online, but it did not have high-definition video. I asked its designers (my former colleagues) why they did not include HD video, and their reply was "The scientists insisted it wasn't a requirement."

By this time I had moved on from Woods Hole to MIT, and for the next six years I served on the committee that oversees *Alvin* and *Jason* for its federal sponsors, playing some combination of devil's advocate and sociologist. I found the scientists' thinking pretty fuzzy, so I started to ask questions to help clarify the debate. If the manned submersible was retained because of the quality of its imagery, then did that mean one should need to pass an eye exam before being allowed to dive? No, that wasn't it, they insisted. "It's the focus," one of them said. "We all know how distracting and busy the *Jason* control van can be; in *Alvin* it's just the two scientists and the pilot and they can really concentrate."

Was it really mental concentration that should justify a new sub? The scientists were trying to articulate something about the quality of their personal experience.

As with other types of remote robotics, the rationale for unique, human abilities requiring direct presence on the seafloor kept shifting as technology evolved. Finally one scientist stood up and said, "We all know that the opportunity to dive in *Alvin* was one of the things that attracted us to this field as graduate students. I'd hate to see *our* graduate students not have that opportunity."

Now we're getting somewhere, I thought. This argument is about professional identity—we are field scientists, we go into the field, and our field is the deep ocean. We need to inspire and train the next generation. Then came the cultural argument: *Alvin* is an icon, people associate it with the institution, it draws people to ocean science and to science in general. Another perfectly good argument—similar to the "spaceflight as inspiration" argument offered to support NASA human spaceflight.

Real resources were at stake; the entire oceanography budget of the United States was roughly $100 million, and a new *Alvin* was estimated to cost at least $40 million, nearly half of it. What's more, the costs of robotic vehicles ranged from a few hundred thousand to a few million dollars, so a new *Alvin* was going to cost the equivalent of forty to a hundred new vehicles. You could cover a lot of ocean with that many robots. But that math only makes any sense if vehicle funding is a zero-sum game. If *Alvin* could secure funding from Congress at levels that no other vehicle could, then of course you'd build a new one—why turn down available funds? Whereas if the new *Alvin* funding would take away from other funding, then there was a serious debate to be had. But of course, this was a question about Washington science policy, not about technical capability.

As in space, warfare, aviation, and countless other domains faced with new robotic ways of working, the most heated arguments are less often about specific technical capabilities than about pride, culture, and professional identity. In the end, the committee—dominated by *Alvin*-using scientists—recommended a new *Alvin* and Woods Hole decided to procure a new pressure sphere for *Alvin*, with bigger windows, but retain most of the rest of the submarine. *Alvin* would now be able to dive to 6,000 meters, not the 10,000 originally called for by scientists.

What emerged from overhaul was a modestly upgraded *Alvin*—the view out the portholes now overlapped, and the viewports were bigger. Also, the new *Alvin* incorporated a host of electronics and software originally developed for *Jason* and autonomous robots.

Thus the remote mode of exploration has not superseded direct human presence. Each redefines the other. Early on, *Jason* had a crude manipulator arm, so *Alvin* defined itself as a vehicle with finesse and strength. When the same arms as *Alvin* appeared on *Jason II*, scientists distinguished *Alvin* by its quality of vision, even when they had not requested HD video on the new *Jason*. All the while pieces of robotics were making their way into *Alvin's* sphere. It now has a Doppler sonar, the same used to navigate ROVs and AUVs, and *Jason's* control software to enable it to run precise surveys and track lines.

How is today's *Alvin* not a robot that people sit inside? Fiber-optic cables are now available in hair-thin sizes and can carry gigabits of data to the surface. Why not connect one to *Alvin* and let the science party on the surface ship join the dives in real time? This idea has been suggested for years, but it has never been done. Would it erase the "autonomy" of an *Alvin* dive? *Alvin* pilots, and their scientist partners, jealously guard the camaraderie and focused isolation of their titanium sphere.

Around the turn of the twenty-first century, whole new classes of vehicles came online. These AUVs (autonomous underwater vehicles) travel under their own battery power, with no heavy cable linking them to the ship. They could be smaller, lighter, and more nimble than vehicles like *Jason*, and could fly ever more precisely due to their complete freedom from the ship. Yet rarely did they swim entirely on their own— whenever possible, they linked to the ship through acoustic communications; even with low bandwidth, these systems could send the equivalent of a text message every few minutes to report on status and position. Like *Jason*, they worked best when they did what they were told and, like *Jason*, their real contributions involved digitizing the seafloor. We'll explore these vehicles' "autonomy" more in Chapter 6.

Meanwhile, still newer vehicles are emerging that blur the lines between human, remote, and autonomous. In 2009 a Woods Hole vehicle broke the record for the deepest dive ever, to more than 11,000 meters in the Mariana Trench. It was not a reconstituted *Alvin*, nor a traditional ROV like *Jason*, but rather an HROV or "hybrid" ROV called *Nereus*, that can operate in a remote (cabled) or autonomous (untethered) mode. And like *Jason*, *Nereus* works off a surface ship, and so is never more than a few miles from its human operators. Sadly, in 2014 *Nereus* was lost at the Kermadec Trench off New Zealand, probably crushed by the immense pressure at its working depth of 10,000 meters (more than six miles). No one was hurt.

Remote presence on the seafloor seems the provenance of nerds as much as great explorers—the exploring body does nothing so strenuous as sitting in front of a computer screen. The data coming in at first seem ethereal, anticlimactic. It takes a leap of imagination to immerse oneself in digital photo mosaics or bathymetric maps. But each is a doorway to a new world, one not imaginable from inside the *Alvin* sphere. Here one can explore at great leisure, over many months, and discuss with colleagues near and far. Here one can float above an ancient shipwreck from any position, see with any perspective.

At one level, all we have done is changed the place where the people are when they do this work. Yet in so doing we have changed the nature of the work, and hence changed what it means to be an oceanographer, an archaeologist, an explorer.

CHAPTER 3

Air

WHEN THE PILOTS OF AIR FRANCE 447 WERE STRUGGLING TO control their airplane, falling ten thousand feet per minute through a black sky, pilot David Robert exclaimed in desperation, "We lost all control of the airplane, we don't understand anything, we've tried everything!" At that moment, in a tragic irony, they were actually flying a perfectly good airplane. The icing problems on the pitot tubes, which had caused the autopilot to trip off and the fly-by-wire system to end its protection of the aircraft, had actually cleared themselves after about a minute.

Yet the combination of startle, confusion, at least nineteen warning and caution messages, inconsistent information, and lack of recent experience hand flying the aircraft led the crew to enter a dangerous stall. Recovery was possible, using the old technique for unreliable airspeed—lower the pitch angle of the nose, keep the wings level, and the airplane will fly as predicted—but the crew could not make sense of the situation to see their way out of it. The accident report called it "total loss of cognitive control of the situation."

Transcripts of the final minutes of Air France 447 make terrifying reading. Pilots feel their awareness disintegrate around them, plummet through the sky, and lose their own lives and those of the passengers in their care. Those four and a half minutes enact, in sad, compressed microcosm, the dilemma of automated systems that we all face: as computerized controls distance us from the direct nature of tasks, we can lose our abilities to work without them, and become different people. When those systems fail, or check out, as they inevitably, if rarely, will, we may be unable to reconnect, and to become again the people we were before.

The world aloft has always been hostile to unaided humans: as Icarus discovered, lose your wings and you plummet to earth. Today airliners travel miles above the earth in rarefied air at high speeds where even a skilled pilot has trouble flying by hand. Lose your computers, it seems, and you may plummet to earth as well.

Unlike an oceanographic vessel, or a robot in the deep ocean, a commercial airline flight deck, while a complex technical system operating in an extreme environment, has direct bearing on our own lives. Each time we walk onboard an airliner, as millions of people do every day on tens of thousands of flights, we entrust our physical safety to the people and machines on the flight deck to get us safely into the air and back down again. What role do the pilots play in getting us there? Given that they sometimes make mistakes, should we eliminate them altogether?

Qantas Airlines pilot Richard de Crespigny has the classic pilot's pedigree. Following a childhood fascination with aviation, and military training, he entered the airline world and worked his way up from the smaller jets to the 747s and finally to captaining the big Airbuses. But in his spare time, he is a self-described geek who programs computers and even operates a small database business. He is also a motorcyclist, someone who feels "comfortable throwing machinery around to see what it [will] do."

In 2004, de Crespigny graduated from the classic series Boeing cockpits with old-fashioned round dials to flying a more highly automated Airbus. "I'd be dumping the Boeing philosophies I'd learned over the preceding eighteen years," de Crespigny wrote, "and replacing them with Airbus's entirely new philosophy, almost tantamount to learning how to fly again." He liked the highly automated Airbus, but it still made him slightly uncomfortable. "I don't like controlling machinery I don't fully understand, a habit formed when pulling apart motorbikes and cars as a teenager. I need to understand the philosophy of how the machine is designed and assembled." In a familiar refrain he concluded, "Airbus pilots are systems operators as much as they are aviators," but still felt he "wanted to go deeper than the Airbus system would allow me."

On November 4, 2010, de Crespigny was forced to go deeper, as he faced a critical situation not unlike the one the Air France crew faced, and in some ways much worse. He was flying the Airbus A380—the giant, four-engine, two-deck jumbo—from Singapore to Australia. Four minutes after takeoff, one of the engines exploded. Shrapnel from a decaying turbine severed myriad fuel, hydraulic, and electrical lines in the aircraft's wing, disabling controls and causing gushing fuel leaks.

The ECAM (electronic centralized aircraft monitor) system—the computer-generated alerts that created so much distraction for the Air France crew—began firing off an overwhelming barrage of messages to de Crespigny. Each message commanded the pilots to execute a checklist, but no sooner had they finished one than another would pop up. "We did one checklist after another, and they kept coming—serious checklists, ugly checklists." In all, 130 minor faults and 120 master caution alarms yelped at them over two hours.

De Crespigny became so overwhelmed responding to the computers' instructions that he could not diagnose the ultimate source of the failures. "It was like being in a military stress experiment . . . We were chasing a computer program around," de Crespigny wrote, "when perhaps we should have been flying the plane and just landing."

Finally, de Crespigny recalled the actions of Flight Controller Gene Kranz during the Apollo 13 emergency: don't focus on your failures, figure out what's working, and work with that for a safe return. De Crespigny marshaled his remaining resources, focused his attention, and the crew landed flight QF32 safely back in Singapore with no injuries.

Every time lives are lost due to human error, we can think of other times when they have been saved by human judgment and skill. QF32, and the "miraculous" 2009 US Airways landing on the Hudson River at the hands of Captain Chesley Sullenberger, seem to show that experienced, skilled, calculating humans are critical safety features of these systems on which our lives depend, the last line of defense when the machines fail.

Air France 447 and others undermine those hopes. In the summer of 2013, pilots of Asiana Airlines flight 214 failed to successfully land their modern Boeing 777 in San Francisco on a perfectly clear day; the crash landing killed three and injured scores. Had these pilots' basic flying skills atrophied in the face of automation, observers asked, only to be needed in a moment of crisis?

Technological change, by its nature, exacerbates these problems. New generations of airliners seem to bristle with greater numbers of black-box safeguards and higher levels of complexity. Digital avionics and software, to be sure, have succeeded in simplifying and improving their interfaces. Their safety record is exemplary, and on balance they have certainly improved safety. But they also add layers of complexity to the operations. As a prime example, the flight management systems (FMSs)—computers that control the overall flight planning tasks—have clunky, 1980s-style keyboard interfaces that force pilots into a process of "syntax, sequence, and procedure, not one of mental imaging."

Aviation is stuck in a paradox: adding more automation, while increasing safety in many circumstances, also places heavy burdens on the pilots. Every technological system can be counted on to fail at some point, and we still keep the humans involved to intervene at these critical moments. But if they are too distanced from the machine, they may lose their skills and have trouble intervening at a moment's notice. This is exactly what happened on Air France 447. Simply blaming the pilots, and attributing these accidents and incidents to "human error" does not capture the essence of the problem.

Researchers have been looking at these questions for decades, and their answers are not simple. John Lauber, a respected aviation safety engineer and longtime member of the National Transportation Safety Board (NTSB), ironically summarized the results this way:

> Cockpit automation increases, decreases, and redistributes workload. It enhances situational awareness, takes pilots out of the loop, increases head-down time, frees the pilot to scan more often, reduces training requirements, increases training requirements, makes a pilot's job easier, increases fatigue, changes the role of the pilot, has not changed the role, makes things less expensive, more expensive, is highly reliable, minimizes human error, leads to error, changes the nature of human error, tunes

out small errors, raises likelihood of gross errors, is desired by pilots, is not trusted, leads to boredom, frees pilot from the mundane, and finally increases air safety and has an adverse effect on safety.

"Automation surprise," "automation dependency," and "automation bias" have all been named and studied by researchers. One pilot I interviewed referred to the computers in his airplane as "a cobra ready to strike. . . . As long as I've got a good long stick and I could poke at it," he added, "I'm probably OK, but I'm not sure that it may actually come back and bite me."

A video from the 1990s depicts an American Airlines training seminar in which an experienced training captain laments the "children of the magenta"—pilots who become so dependent on automation that they sit back and let the computer follow its automatic course, depicted on computer screens by a magenta line.

Still, today most pilots, like de Crespigny, started their careers and training in traditional "round dial" cockpits and then transitioned to computerized ones. But now young airline pilots are making their way through the system having never flown an aircraft with a traditional altimeter or airspeed indicator. Are they different people? As one young pilot put it to me, "It seems that now things have changed . . . it's just a straight job. You go in, you turn the computer on, you get down back on the ground, turn the computer off and go home."

What is this new experience of this century-old profession?

"What's it doing now?" is a question all pilots have asked of their computers at one time or another (researchers call this "mode awareness"). "Every time I say that, I cringe," one pilot says, "because I know right good and well that ninety-nine percent of the time it's doing exactly what I've asked it to do, and it's what I've asked it to do that I'm not understanding." Still another answers that only the inexperienced pilots ask "What's it doing now?" The experienced pilot will just shrug and say "It does that sometimes"—a defeated resignation to a new reality. Others bristle at the complacency: "'Why is it doing that' is not the question. If you have to ask the question, it's 'Why am I letting it take control of me?' Get rid of it. Take the airplane back."

At stake are basic ideas about human agency, control over one's work, and how to live and work within a system. The airline pilots' identity crisis mirrors that of undersea scientists, drone pilots, space explorers, and those in numerous other professions whose cognitive capacities are being pressed by computers and automation to live in a digital and virtual world while still suffering the consequences and literal hard realities of the physical world. In a future of automated cars, driving may resemble the anxious monitoring of airline pilots more than the calm relaxation of passengers along for the ride.

In late 2013, a joint industry–FAA working group on flight deck automation assessed the current state of technology and pilot skill. They reviewed twenty-six recent incidents and accidents and found that pilots may rely too heavily on automated systems, not understand them sufficiently, and make errors while using them. Moreover, the knowledge base and set of skills required of pilots has expanded. While some traditional skills are now "reversionary," to be used only in emergencies or failures, pilots must still be proficient with them. The definition of "normal" pilot skills has changed, and more is expected of today's pilots than even in the recent past.

Kathy Abbott, a PhD computer scientist, is the FAA's chief scientific and technical advisor for flight deck human factors and co-chair of the working group. She points out that automation does not remove human involvement in the operation of the airplane, but rather moves it around. "We're not eliminating human error" by automating tasks, Abbott explains, "we're just changing where the error occurs." Furthermore, she points out, experienced pilots make just as many errors as less experienced pilots, but they make different types of errors. Cognitively, error is "the downside of having a brain," and essential to learning.

Data show that only 10 percent of commercial flights proceed exactly according to the flight plan. Some 20 percent of flights raise some kind of malfunction in the systems that require a response from the crew. Because most of these data come from accidents and incident reporting, Abbott and the working group found, "there are very few data sources that capture the positive aspects of the aviation system," such as the numerous, daily interactions in which pilots make the system work by overcoming, often in invisible ways, errors of both human and machine.

In fact, the data revealed that pilots actually do a great deal of invisible activity that mitigates risk, and that the overall system is designed to rely on that critical role. For example, pilots cannot always follow standard operating procedures, either because those procedures don't perfectly match the situation or because they might be too lengthy or detailed to complete in the time available for them. Few have studied these subtle dimensions of pilot behavior, yet understanding them is an essential prerequisite for conversations about unmanned airliners.

The procedures that pilots use are not a set of ironclad laws, but rather compromises between performance, simplicity, liability, and a host of other factors. They are like computer programs for people—instructions written by other people. Emergency situations frequently do not correspond exactly to those foreseen by engineers and encoded in checklists. As on QF32, pilots have to improvise in critical moments, sometimes against the advice of their computers. Pilots are the glue that hold integrated systems together, making up for imperfections in protocols, communications, interfaces, or procedures.

These findings suggest that new technologies ought to enhance human problem solving, not eliminate it. If possible, automation ought to aid humans in their tasks without distancing them from the machine, and without alienating them from their professions.

Aviation has always been a realm where human identities evolved along with changes in technology. To grasp the current and future questions, then, we need to begin with a little history of the pilot, that central character in the evolution of the modern human, whose experience has been constantly changing alongside the technologies of flight.

Of all of the Wright brothers' contributions, their greatest was simply their idea that an airplane should be a machine under active control of a person. This idea made the airplane practicable, and also generated one of the great social inventions of the twentieth century—the airplane pilot, master of machine, traveler in an unstable element, and surveyor of human life below.

"The twentieth century was born yearning for a new type of hero," writes aviation historian Robert Wohl, "someone able to master the cold, inhuman machines that the nineteenth century bequeathed and at the same type transforming them into resplendent art and myth." From Charles Lindbergh to Neil Armstrong to "Sully" Sullenberger, the cultural icon of the pilot embodied the human on the cutting edge of technology and social change. Analogies flowed freely—the adventurer of the sky, the aerial artist, the athlete of the third dimension. World War

I offered new identities, particularly the "knight of the air" flying fighter ace, reviving ancient mythologies to rescue heroism from an anonymous war of trenches and random death.

Ironically, most pilots, including Wilbur Wright, Charles Lindbergh, and Neil Armstrong conformed to none of these archetypes so much as to that of the mechanic or engineer imbued with the middle-class virtues of temperance, deliberation, and focus. The story of the pilot in the twentieth century is in part the story of the conflict between the dreams people had for pilots, the public mythology, and the actual characteristics required to fly successfully and survive. It is also a story of human identity evolving hand in hand with new technologies.

Even as they reveled in their freedom and autonomy in the skies, pilots have long tempered their joy with laments over their perceived loss of control, intuition, and sensation in exchange for greater stability, safety, and capability. Aviatrix Beryl Markham, writing of her bush flying in Africa, saw the decline of "this era of great pilots" much as the era of great sea captains had disappeared, "each nudged aside by the march of inventive genius, by steel cogs and copper discs and hair-thin wires on white faces that are dumb, but speak."

Markham's lament sounds familiar today, but she was writing in the 1930s. Her "hair-thin wires," the magenta lines of her day, were the new instruments invading the cockpit that told the pilots how to fly straight even while immersed in clouds. She too lamented that pilots would lose their skills: "If you can't fly without looking at your airspeed and your altimeter and your bank-and-turn indicator, then you can't fly."

Markham was responding to the rise of instrument flying, the analog, 1930s equivalent of computers and automation fifty years later. Instrument flight allowed pilots to fly without visual references to the horizon or the ground; they could focus solely on visual indicators inside the cockpit. Jimmy Doolittle, hero of the eponymous "Doolittle Raid" of World War II, actually conducted some of the earliest studies of pilots in their cockpits. He proved that for a pilot flying in clouds, Markham's sentiment was wrong—if she can't see the ground or the horizon, it is actually impossible for a pilot to fly by her senses alone.

The solution was to place in the cockpit instruments driven by clever gyroscopes that always indicate which way is up. Doolittle flew the first instrument flight in 1929, taking off, flying a full circuit around the field, and landing, all with a hood over his cockpit and referring solely to his instruments. Still today, instrument flying requires a special rating from the FAA. Pilots have to be trained to trust their eyes and their numbers more than their own bodies—in the clouds, flying by the "seat of the pants" will kill you in a matter of minutes.

The new technology entailed a change in identity. Pilots could no longer be rough, intuitive mechanics, but had to become numerate, educated rationalists. Charles Stark Draper, who pioneered the design of these devices, intentionally called them "instruments" to give the cockpit the modern authority of the scientific laboratory. The rise of commercial airlines in the 1930s demanded both technical and social progress—airplanes that could fly reliably and safely in all kinds of weather, and clean-cut operators in uniforms who could offer the public assurances of sobriety and comfort. For the flying public to accept commercial flights as routine and safe, they had to perceive the pilots, like the airplanes, as stable and controlled.

Robert Buck exemplified this evolution. His career as a pilot began in the instrument era and ended in the era of automation. A prolific, articulate aviation writer, Buck gave voice to the anxieties of his generation. Early in his career in the 1930s, the modern DC-2 and DC-3 airliners "required the pilot to start down the road to being a technician."

Buck's memoir of the ensuing forty years is a narrative of pilots' loss of control and autonomy. Technology was not the only culprit: for Buck, management, government regulation, and engineers were also to blame.

Often this change is described as a transition from hands-on craft flyers to pilots as "systems managers." "This expression grates on my sensibilities," Buck wrote. Whatever is going on, "you'd better have a lot of flying knowledge, background, and judgment, which makes you a pilot, not a systems manager."

Writing in 1994, Buck described flying the new "glass cockpits" endowed with computer displays as "a little like entering a modern art museum" with colors pleasing to the eye that "look more like an abstract work of art by Mondrian or Davis than the traditional airplane's instruments." Cognitively, flying became increasingly a matter of seeing and responding to visual stimuli.

Buck's invocation of modernist painters is apt, for the fragmented visual field also symbolized a fragmented professional identity. For Buck, all of this added up to "the pilot's burden," as he was now responsible for "more tasks, greater responsibility, and the necessity for diversified knowledge that includes science, law, and psychology." Buck did not resist cultivating this "broad base of intelligence," but noted that he would be called on to use it all in an emergency, "without the luxury of time for contemplation."

Buck's observations, and those of countless other pilots, underscore how the flight deck has always been a contentious space wherein technology and labor intersect much as they do elsewhere in modern life, but with higher stakes and perhaps greater intensity. Pilots are white-collar workers who manage teams of people operating large, complex systems in high-risk environments. Yet they are also craftsmen who rely on their manual skills and form labor unions.

The language we use for members of a flight crew suggests the ongoing adjustments of their roles. In the 1930s, Pan American World Airways (Pan Am) began to replace the terms "pilot" and "copilot" with "captain" and "first officer," and gave them the now-familiar maritime-inspired uniforms to suggest confidence and authority based on established social roles. More recently, these terms morphed into "pilot flying" and "pilot not flying," because the captain might not always be the person flying (or, as on Air France flight 447, the captain may not even be in the room). Now, the Federal Aviation Administration (FAA) has recommended these terms be changed to "pilot flying" and "pilot monitoring" to give positive designation to their actions, showing both pilots are engaged in flying the plane regardless of which has a hand on the controls. (In these conversations "flying" often still refers to hands on the controls, even though "flying" overall encompasses many other activities.)

Terminology is not the only change. As we look at current changes in pilots' roles, we find one dramatic precedent: the elimination of the flight engineer, the cockpit's "third man." In the 1920s, commercial aircraft were designed for two pilots. As aviation evolved and flights went farther afield, airlines began sending licensed mechanics along to repair aircraft on the ground because they were often flying to places without adequate infrastructure for repairs. As propeller aircraft became two-, three-, and four-engined, each adding multiple dials and controls, simply managing the engines became a major part of flying the aircraft, and the flight engineer evolved into a professional specialty.

After World War II, the U.S. government began requiring professional flight engineers on commercial aircraft above eighty thousand pounds, which included most of the day's large airliners. Ironically, some aircraft, such as the DC-6, had already been designed for a two-person cockpit when the rule was imposed. Yet the third man was still required, even though he had little to do. He therefore rode in the jump seat between the pilots, and helped provide an additional pair of eyes to avoid conflicting traffic.

There followed a period of conflict between pilots and flight engineers. Increasingly, airlines put professional pilots in this third seat; pilots' unions refused to accept flight engineers as members. Flight engineers countered by arguing that the pilots themselves had little to do during the bulk of the flight. During congressional hearings they presented photographs of pilots reading newspapers, sleeping, or flirting with female flight attendants sitting on their laps while en route.

Again, technology changed the roles and relationships: jet engines, which appeared on commercial aircraft in the late 1950s, had greatly simplified mechanisms and controls—"dozens of knobs, dials, and gauges vanished from the cockpit." A study by the British Airline Pilots Association found that "very little, if any opportunity exists for the use of specialized mechanical skill in flight." Manufacturers believed that the first-generation jets, the Boeing 707 and the Douglas DC-8, could be operated safely by two pilots, but they were required by the FAA to include the third man (the military version of the 707, the KC-135, a nearly identical aircraft, had no flight engineer). Statistical studies about whether the third man improved safety were inconclusive.

In 1980, President Ronald Reagan appointed a panel to study the problem. They concluded that jet transports with two pilots were safe, that the third pilot did not increase safety, and that the new Boeing 757, 767, and Airbus A-310, then on the drawing boards, could be safely operated by two pilots because of their computerized, "glass" cockpits. The eighty-thousand-pound rule was rescinded in an era of deregulation and anticipation of new electronics.

The story of the third man is surprisingly relevant for today's pilots.

Indeed, the challenges they often face—like de Crespigny's troubleshooting failures in complex systems—were once the job of the now absent flight engineer. What used to be conflicts between people on the flight deck—the pilots and engineers—have now become conflicts between people and computers.

Yet when we ask "Where are the people?" we see how each computer still embodies human effort. And the work embedded in those machines, like the people themselves, exists in imperfect relationships. The instruments and dials on the flight deck are often made by different vendors and programmed by disparate software teams, yet all have to work together as a single system.

Draper intended his term "instruments" to refer to science, but "instruments" has a meaning in music too: in that sense, pilots conduct the orchestra, keeping the instruments together in concert, easing the constant, relentless, and unforeseen mismatches between the players. Adding new instruments may change the sound of the music, but how does it change the job of the conductor?

When interviewing pilots for this book, I asked them, "What has changed the most about flying during your career?" I expected them to respond with a comment about computers or automation. Instead, most pilots answered that the number of their responsibilities, especially on the ground, has grown. In an earlier era, pilots might walk up to the plane, inspect it, and get in and start flying. Dedicated gate agents would arrange fueling, catering, and the numerous other services an aircraft requires on the ground.

Now airlines often cut back on those jobs, leaving more responsibilities for the pilot before the engines are even started. "We work more. We fly more. Our days get longer." Ordinary technology abets the change: each cockpit now has its own mobile phone, with its own designated

number. The pilots are responsible for communicating with maintenance, dispatch, and other airline services, reducing their autonomy.
"You always have somebody in your neck. Five people are calling you by
phone. The gate called. The flight manager called. The company called.
What's going on? How long do you need? Is somebody there?"

By the time they fasten their seat belts and begin operating the aircraft, the pilots are already well into their workday. Moreover, once they
are in the air, they have to take on increasing requirements for noise
abatement, fuel efficiency, speed to the gate, and traffic controls. Picture
the way your own experience of air travel may have changed in recent
years; the flight crews' jobs have changed as well.

Although immersed in an extreme environment, the flight deck
remains a workplace where strong professional identities meet rapid
technological change, where the autonomy of the skies meets the eco-
nomically stretched air transport system, and where the individual's
command and responsibility for human life must work within high
degrees of government regulation. When we think about automation
and technology aboard airliners, we must imagine them within this
ever-changing environment. The children of the magenta line are not
flying through the same world that Beryl Markham and Robert Buck
did. These forces often converge during the critical moments of landing.

Every flight is a story, and the climactic moment is the landing, which
has always been a focus of piloting skill. It is the most difficult skill for
student pilots to learn, and good pilots challenge themselves to land
perfectly every time. Weather makes the problem harder.

In bad weather or poor visibility, pilots fly on instruments, looking
down at the glowing array of dials and numbers that tells them where
they are, how the plane is flying, and how they're doing. During landing,

this environment epitomizes everything you are trained *not* to do in every other phase of flight: flying low and slow, flying close to the ground, and not seeing anything.

If you've ever looked out the window of an airliner while it was landing and seen nothing but white cloud as you felt the landing gear drop and heard the engines spool down, then seen hard ground appear just as the plane touched down on a previously invisible runway, then you have some notion of "minimums."

As pilots descend through clouds or overcast, they look through the window for signs of the runway: flashing strobes, black pavement, or special cross patterns of ground lights designed to mimic to the pilot's eye the instruments in the prior glance. If any of these appear, the pilot can go visual, follow the eyes, and land by sight.

Minimums are also called the "decision height"—usually about two hundred feet above the runway with about a quarter mile of visibility. Below that, without seeing the runway environment, a landing is illegal and unsafe and the pilot must push the throttles forward, fly away and hope to try again, or go somewhere else. Flying an approach to minimums takes confidence, experience, delicate and well-tuned instruments, and a great deal of trust in the system—accurate procedures, well-maintained equipment, competent air traffic control. For passengers, landings in such conditions enable us to catch our flights on rainy days. For airlines, instrument landings are money in the bank.

Traditionally, aircraft fly using an instrument landing system, or ILS, a radio beam that emanates from the runway and provides a right/ left error signal (the localizer) and a vertical beam (the glide slope). This system has been in place since at least the 1950s, and allows an aircraft to literally fly down the beam to as low as 200 feet before the pilot sees the runway. This is known as a Category I landing (abbreviated "CAT I" in the lingo). With more advanced equipment operating to higher precision, such as a radar altimeter to directly sense the ground, pilots can fly as low as 100 feet, a Category II landing.

Going further, Category III minimums are so low that a human pilot alone cannot make the decision and must rely on some form of automation. Category III has been divided into smaller bins, where CAT IIIa has a decision height of 50 to 100 feet, and CAT IIIb of 0 to 50 feet. CAT IIIc is total whiteout, "zero-zero"—a cloud ceiling that goes all the way to the ground with zero horizontal visibility. CAT IIIc has no decision height, and while some aircraft (and very few airports) are certified to these conditions, they would be unable to taxi off the runway, so it is rarely used in practice. The minute differences between landing categories can seem a bit arcane, but for our purposes it's just worth remembering that CAT III landings can be extremely challenging.

One way of handling Category III is via automatic landing systems, or "autoland," first brought to operational maturity in the 1960s in northern Europe. British Airways found that fog was affecting 7 percent of their flights at their London Heathrow hub. Combining the standard ILS radio beam and onboard inertial guidance, autoland can bring an airliner down with zero visibility, automatically raise the nose (or "flare" the aircraft) over the runway, and command automatic braking to bring it to a stop. Where a standard ILS gets a pilot to 200 feet before having to make the landing decision, autoland can get down to zero, or CAT IIIc. "Look! No Hands!" wrote pilot Richard Collins when he test-flew autoland in the 1980s.

Autoland seems like an ideal solution, the perfect backup to the human pilots on a difficult day. Today, most Boeing and Airbus aircraft have autoland as standard equipment (for aircraft with digital fly-by-wire flight controls, autoland is a relatively simple addition to existing systems and software).

Yet despite its name, autoland can be complex to operate. It requires double or triple redundant autopilots running off different electrical systems, and has limitations regarding wind and inoperative equipment—everything has to be working perfectly, in only moderate headwinds and gusts, for autoland to work. Moreover, autoland is not just a box on an airplane, but requires infrastructure on the ground to operate at a high level, and crews and even the airlines' organizations to be trained and certified in the operation. In the United States, high winds are often associated with low visibility, rendering autoland unusable. But in fog or low overcast with low wind, it can mean the difference between getting in and diverting—even in most demanding "zero-zero" conditions.

As with all automation, autoland does not replace the crew altogether. For one thing, they have to decide when to turn it on. "The captain is in charge, monitoring, making decisions," Collins wrote. "He is involved but detached." The specially trained crew has to set the system up, keep an eye out for failures, and take over in case of a problem. To be ready to intervene in the case of failure, they may also keep their hands on the controls while the autoland flies. Richard de Crespigny recalled the autoland on his old Boeing 747 Classic as "a humble mechanical device, incorporating lots of servos and actuators that delivered mediocre performance and reliability, and needed to be checked regularly."

Modern autoland systems are reliable digital boxes. But if there is a failure during the critical moments of landing, the pilot must employ a sequence of logic and choose a set of actions, such as land manually, command automatic go-around, or go around manually. Because autoland demands an extremely high level of precision from ground-based systems, aircraft on the ground have to be clear of the runway and adjacent taxiways in order to not interfere with the radio beams, which means airport capacity can go down by as much as 50 percent during autoland operations. At least one airline found that pilots were using autoland on only about 2 percent of landings, and those were mostly to keep the crew and the equipment up to minimum certification.

Nevertheless, autoland is impressive in its precision and safety. Collins found that autoland might actually raise the standard for human performance: "Watching the precise performance is enough to make any pilot want to rise to this computerized electromechanical challenge of skill." He found that watching it fly was "a lesson in how to land an airplane properly." Nevertheless, according to the 2013 FAA working group on automation, "the circumstances requiring or allowing an automatic landing are rare, and pilots normally prefer to manually land the airplane."

Autoland might point toward completely automated, unmanned airliners in the future. But as we saw in the world of deep-ocean robotics, these technologies do not necessarily move in linear progression from manned to autonomous. Other solutions now exist for integrating pilots more deeply into the control loops. The heads-up display (HUD) fuses a computer-generated virtual world with what the pilot sees through the window, enabling the combination of human and computer to do more than either could do alone.

It's worth taking a closer look at HUDs because they seem to offer a new approach to the pilots' roles, and it is an instructive one. They demonstrate how it can take newer, more sophisticated technology to include the person in less automated, more precisely defined roles.

On a clear spring day I'm sitting in the jump seat of a new Embraer

190 jet, behind and between the pilot and copilot on a commercial flight. On our way to Geneva, Switzerland, we are flying southwest across Germany along the northern edge of the Alps. From 28,000 feet, the view is stupendous. We pass over the fantastic Neuschwanstein Castle, model for the Disney castles, gleaming at the foot of the mountains. Then over the Black Forest and Lake Constance, where the Hindenburg was built. Mont Blanc looms in the distance.

This Embraer is a twin jet, similar in form to the Boeings or Airbuses that cross the oceans, but smaller. It is a regional jet, designed to

Airline captain using a heads-up display, flying over the Alps on an approach to Vienna. Note the glass "combiner" that displays the heads-up image in the pilot's field of view.

(PHOTO BY THE AUTHOR)

replace older jets and propeller planes on relatively short trips, adding all the modern conveniences and safety features.

We sit in a fly-by-wire plane under full digital control. Five page-sized computer screens, two on each side and one shared in the middle, sit in front of the crew. Each crew member has a keypad on the center console for typing into the flight management computer, as well as a shared pair of throttles for the two engines. Each pilot has a control yoke, a gull-wing-shaped handle that's an Embraer trademark. The dizzying array of dials and indicators of yesteryear is gone; the look is clean and well-thought-out, like that of a modern office.

It is late in the day and the pilots are tired. As we prepare to begin our descent into Geneva, the flight attendant brings in little cakes for a snack—a little blood-sugar boost for landing. The pilots put their shoulder belts on in preparation for the arrival.

Yet as the plane approaches the airport, the pilot does not sit back and monitor the autopilot as it flies down the radio beam, nor does he switch on autoland. Rather, the pilot is hand flying, his gaze pegged to a small piece of glass hanging down in front of his eyes from the ceiling. As he looks through the glass, numbers appear on the periphery of his vision, vital statistics like airspeed and altitude. In the center of his view, a computer image outlines the runway in glowing green. Today he can see it clearly, but in bad weather the virtual runway would overlay the fog and the pilot could land as he normally would.

A little symbol on the display is shaped like an aircraft and called the "flight path vector." It tells the pilot where the airplane is going. A second icon, the "guidance cue," a little circle, tells him where it *should* be going. With his hands on the yoke, the pilot flies to keep these two symbols together. As we smoothly approach the runway, the guidance

View through a modern heads-up display (HUD) on an approach to an airport. Note how the flight path vector (the small circle) is lined up with the approach end of the runway, as well as a dashed line indicating the appropriate descent angle. The traditional "glass cockpit" primary flight display (PFD) is below.

(PHOTO BY THE AUTHOR IN A FLIGHT SIMULATOR)

cue changes from a circle into a cross and begins to rise. The pilot pulls back on the yoke to keep his flight path vector centered on the cue, hence flaring the airplane as it slows down. It kisses the runway, a nearly perfect landing. All the pilot had to do was keep the two green icons centered, "keep the thing on the thing," and his smooth touchdown was guaranteed. Had the airport been fogged in and the runway obscured, he would have done nothing differently.

Heads-up displays have been on military aircraft for a long time. They originated as gun sights in World War II that presented a reticle sight to a pilot that automatically corrected for the lead of the target. On modern fighters they have become advanced sights that present critical flight data, identify radar targets, and aid the pilot's aiming.

In the 1980s HUDs began appearing on commercial aircraft. The original ones presented flight data from the instruments (airspeed, altitude, etc.) so the pilot wouldn't have to look down while in the high-traffic airport area and during critical phases of landing. These devices, including many still flying on commercial airliners today, had relatively narrow fields of view, forcing the pilot to sit in a precise position and look through a small window—a position pilots often found uncomfortably constraining.

Newer HUDs have larger displays and are "conformal," that is to say, if the pilot holds his or her head in the right place (by aligning the eyes according to a set of fixed markers on the instrument panel), then the images on the HUD appear overlaid exactly where they would be in the real world. For example, if a pilot is landing in weather with low visibility, the HUD actually paints the outlines of the runway in its green computer graphic vectors; the runway shape will get bigger as the aircraft approaches, and then when the runway finally appears out of the clouds it sits precisely behind the green box.

Eventually, HUDs were certified by the FAA and European authorities for Category III landings. The US Airline Pilots Association, the leading pilots' union, supported them as preferable to autoland. Alaska Airlines, which flies difficult approaches into Juneau, became a lead user; a small airline called Morris Air also began experimenting with the system. When Morris Air was acquired by Southwest Airlines in

1992, Southwest adopted HUDs for much of its fleet. Pilots came to rely on the HUD for low-visibility landings and would therefore sometimes ignore or disconnect the autoland systems.

Today's HUDs have their own guidance algorithms, which take navigational data and present the HUD's own predictions about where the aircraft is heading. The flight path vector icon incorporates everything the computer knows about the flight, including wind speeds, inertia, and power settings on the engines. The pilot can move the flight path vector by moving the controls (and hence moving the airplane's direction of flight). If the pilot flies so as to place the flight path vector on the end of the runway and keeps the icons aligned, then the aircraft will fly right to that point—and if it's at the right speed and altitude at that moment, it will land with only gentle inputs from the pilot. The subtlety of manual skill now responds to the visual task of "putting the thing on the thing."

The HUD also has symbols, "error tapes" that help the pilot keep the aircraft at the right speed and acceleration. With traditional flight instruments, the pilot sees airspeed, altitude, and vertical speed—leaving to mental computation whether the aircraft will fly the right path. By contrast, the HUD allows the pilot to see the aircraft's "energy state" directly—and literally as well: will there be enough momentum to make the runway? These indicators change more quickly and accurately in response to small errors, allowing the pilots to make small corrections before it's too late.

Another symbol previously mentioned, the guidance cue, even tells the pilot where to put the flight path vector. Again, all the pilot has to do is manipulate the controls to "put the thing on the thing"—follow the guidance cue with the flight path vector to make the aircraft fly the right path.

These new, digitally synthesized instruments gradually led pilots to see that the HUD might be useful beyond low-visibility landings. One HUD engineer recalled, "Pilots started saying, 'Whoa, if I can use it for this, if you just made this other change and made it look like my [traditional] primary flight display does, then I could use it all the time.'"

In my research I interviewed a number of pilots whose airline (I'll call them EuroAir) was choosing HUDs over autoland for a new fleet of Embraer jets. Some of the pilots had come from aircraft with older HUDs, and some had come from autoland-equipped aircraft, so the debate at the airline over the new jets had two natural constituencies.

Pilots live in their aircraft, and I rarely found them criticizing their workplace—in general they make peace with the airplanes they are assigned to fly, and use their features to the greatest extent practicable. Similarly, pilots assigned to train in a newer aircraft were unlikely to be critical—the Embraer represented the future, fly-by-wire technology for the airline, and pilots would be foolish to reject it. Some even felt that training in the Embraer would give them an added measure of job security, given economic uncertainties and possible future mergers at the airline. Nonetheless, other pilots were frank about their hesitations in losing autoland.

Officially, the rules for autoland require just as much pilot vigilance as hand flying, but pilots admitted it wasn't always used that way. "When you're tired in the morning, just do autoland." Nearly every pilot mentioned the long days and fatigue, "days when you have five legs, you have ten hours flight time, or even twelve hours, and the last leg is a CAT III approach . . . and there, autoland was really something." But too tired to hand fly should also mean too tired to monitor.

The pilots who had flown HUDs before were not necessarily more eager to fly HUDs in the Embraer. The older-style display they were used to had a rather small "combiner" (the glass window the pilots look through), so the pilots had to adjust their seat position to an awkward spot in order to use it. Moreover, it was primarily used for CAT III landings, "so we ended up actually using the system with a low level of proficiency under the worst weather conditions, with a rather uncomfortable seat position."

There were two crucial innovations in the newer Embraer HUD installation. First, as I've mentioned, the new combiner windows were larger, making for more comfortable seating position. Perhaps more important, the new Embraer had HUDs for both pilot and copilot, whereas in the old system only the pilot had one. This simple difference, though twice as expensive, had important effects on the crew's use of the system.

Thomas is an advocate for HUDs at the airline, but he found himself confronting "the old perception that heads-up is a tool for the poor, regional airlines to make their aircraft CAT III capable." So he changed his rhetoric: they began describing the HUD as "a flight safety feature . . . [that] integrated the pilot with the aircraft again" rather than calling it a low-visibility tool. He played on a reaction among pilots to what they perceived as overly automated Airbus aircraft.

Here he was aided by the manufacturer, which had seen a similar progression. The airlines initially bought HUDs for their low-visibility approaches, but ended up keeping them for other reasons. The American maker of the HUDs, we'll call them HudView, found that CAT III landing was a pretty narrow basis for their business. In the United States, there are only about a dozen airports certified for CAT III approaches; it's a niche application. The pilot could flip the HUD up to the ceiling and out of the way if so desired, but HudView was finding that pilots were using the HUD throughout the flight. "And I want two of them," the customers were saying. "I want the pilot not flying using it, I want him to be aware of what's going on." Double the business.

Hank is a man in his sixties, just about to retire from HudView when I interviewed him. He had a background in mathematics and statistics, and had done crucial studies to prove the HUD's safety for the FAA. He began comparing HUD-enabled landings to non-HUD landings, over thousands of cases, and began to notice something: "Dumb things that keep happening like tail strike and like hard landings. And I took maintenance rates for landing gears, and wheels, and things like that." After talking with customers and manufacturers, he found that "what's really going on here is they put the HUD on the airplane, and it reduces the cost of [maintaining] the airplane," potentially by millions of dollars per year over a large fleet.

Softer, more uniform landings meant lower maintenance costs, less wear on tires, fewer tail strikes (where the rear of the aircraft scrapes across the runway). In addition, he found the pilots using the HUD in all weather. Because the good-weather landings and the low-visibility procedures were the same with the HUD, "they build up trust in the system as though they were making the low-visibility flight day in and day out." The social idea of trust captures the idea that the person finds the machine predicable and reliable.

Some airlines were ordering HUDs without even the CAT III capability installed, which felt odd to Hank because to him CAT III was what made it special. But the airlines just wanted the added awareness: "I'm looking out the window and I know what's going on" [the pilots were saying], "flight path [vector] tells me where I'm going as opposed to where I'm pointed."

"That was not what I expected for many, many, years," Hank added, "to the point where I'm still sort of surprised by the whole thing." As so often happens with automation, users were applying the technology in innovative ways.

Hank did a study comparing HUD landings to those with autoland. "It's pretty amazing," he said of the statistics. "We found out it works at least as well, and depending on your opinion, it may work better." Hud-View was careful, however, not to anger the airplane manufacturers who were selling autoland with their big airplanes, so the company never published the study.

"The autoland does some things pretty suddenly and strangely," Hank observed, because "it's an automation system." In sudden gusts of wind or turbulence, "the reaction of the autoland is completely different from the reaction of a human . . . [and] can put the airplane in the grass pretty easily in the case of turbulence."

Hank found that pilots didn't trust their autoland systems. "They just didn't know what was going on there, this thing's flying my airplane, I have no insight into what's going on." When equipped with the HUD, Hank found, "they started to think, OK, now I'm flying the airplane again." Some customers started to ask for the HUD instead of autoland.

Hank began to see the HUD as a kind of hybrid automation, as opposed to the "pure" automation of autoland. When customers started to see HUDs in this way, "it seemed like a huge leap of trust in our favor when that happened. . . . We had pilots saying I'd much rather fly this because I know what's going on versus an autoland system." One engineer describes the HUD as a mix of a traditional flight display and autoland, "and the part in the middle is that there's no actuator, or the human is the actuator."

Thirty years ago Richard Collins wrote that pilots, by observing the uniform precision of autoland touchdowns, would improve their performance. Now they are saying the same thing about HUDs—being back "in the loop" improves their skills through hands-on practice, not just by example. As some pilots say: "It made me the pilot I always thought I was."

Some data suggest HUDs can reduce stress and strain during critical phases of flight. One study found that the pilot's stresses when using autoland were quite high at the critical moments; by contrast, stress with a HUD was higher, on average, but had less extreme peaks during landings.

Tom is a pilot in his fifties who works for the FAA, checking out pilots to certify them on large jets. As an instructor in highly automated aircraft, he sees pilots focusing too much on the computers, too much head-down time and not enough "flying the airplane." He sees the HUD as a kind of missing link. "It's covering a step that's in between that we should have covered before. I don't want to say it's a step backwards in automation, but I think it's a gap that we left when we jumped [ahead] in automation." For him the HUD undermines the automation myth of progress—the idea that somehow pushing people out of the loop is inherently more advanced.

John, an instructor on an American airline that also uses dual HUDs, goes further, describing the HUD as "a different way to interact with the airplane, especially when you're in visual environments." He believes pilots with HUDs can fly more smoothly and have higher levels of alertness than pilots without them.

Thomas, the HUD advocate at EuroAir, believes the HUD represents a profound shift. "We are entering a new way of flying with flight path and energy." The traditional way of flying an aircraft is with "pitch and power"—for any given maneuver, set the nose at a given pitch, apply a predetermined power setting to the engines, and the airplane will fly in a well-understood way (one way the Air France crew might have gotten themselves out of trouble). By contrast, with a HUD pilots fly with the flight path vector and the energy vector, what Thomas calls a new philosophy: flying with flight path and energy. Changing the data display actually changes the skills required, returning the flying task to its continuous, visual roots. "Since the HUD is flight path–centric," another pilot comments, "all the information and cueing from the mental side is here, which serves to reinforce one way or the other the physical cues."

The most important safety benefits, however, may be intangible. As an airliner descends out of the clouds toward the runway, the pilot no longer has to look "heads down" at the instruments and "heads up" to search for the runway lights, and no longer has to make the cognitively difficult transition from instruments to the sight of the runway, a known source of risk. What's more, with a HUD every single landing requires the same exact procedure, whether on a perfectly clear day or on a tricky landing in rain and fog. The real-time monitoring of the aircraft's energy state quickly shows whether the airplane is above or below where it's supposed to be. Pilots can now manually fly Category III approaches.

From my own observations, pilots using a HUD are highly focused; almost like someone staring down into a smartphone during conversation. If you are in the copilot's seat and don't have access to the HUD data, the lack of information can feel alienating; you might feel confused about what's going on. The unequal distribution of information exacerbates the existing power differential between pilot and copilot.

Without a second HUD, the copilot is left to monitor traditional instrument and aircraft systems with less information with which to intervene in case of a failure; with a dual HUD, both pilots have access to all the information, and can check each other's performance.

A practical advantage of the HUD is that it enables the pilots to fly every approach, every day, exactly the same way. No more changing of procedures for Category I to Category III, much less the different equipment with different operations, maintenance, and proficiency requirements. "After a while you don't even notice if it's CAT III . . . because it's every time the same and you feel very comfortable," one pilot observes. "This makes a flight more precise," says another. "It feeds back the position of your flight trajectory directly to your eye."

Pilots go through several phases of adapting to the HUD. In initial training, they focus on the HUD data to the exclusion of the view outside the window, what some call "tunnel vision." Optically, in one pilot's words, "while HUD is focused at infinity, your mental focus can be easily drawn to three inches from your face." After training, however, they gradually learn to take in all the data being presented and merge it with the visual scene into an integrated picture. "I don't find myself consciously directing my eyes around the HUD like I do when I'm looking at round dials," one pilot experienced with HUDs observed. "I'm sure that's taking place, but it's not as much of a conscious effort." The process of adaptation can take anywhere from six months to a year of regular flying.

New skills that rely on new equipment, of course, give rise to the worry that pilots will lose the ability to fly without it, leading some to worry about "HUD cripples." Some pilots suggest they should regularly fly visual approaches without the HUD to maintain their proficiency.

Other pilots object to HUD on principle; one instructor recalled a pilot who called it "that little piece of crap that sits in front of you," because he felt it was second-guessing his skills. Some pilots wonder why the airline puts money into technology when they could put it into pilots' paychecks. "Save this money and pay me more . . . I didn't need it before, why do I need it now?" Another reaction was simply "I don't trust it" (to which an engineer would respond, "But you trust your autopilot?").

More common objections are to the mundane practicalities of the system. One pilot noted that when turning off the HUD, "it feels like taking off a T-shirt that's too tight," because of the way it constrains his vision and body. Others agreed it was "nice to have" but not a panacea for all risk problems. Some objected to airlines mandating use of the HUD during all phases of flight. I observed small acts of resistance— the airline monitors whether the HUD is on all the time, but doesn't monitor the brightness of the display. By simply turning down the brightness to zero, pilots can disable the HUD without management's knowledge.

Amidst all this complexity, and all these varying opinions, nearly all pilots I interviewed agreed with the benefits of being "in a position as the active controller" and "in the loop." Some felt flying with the HUD enhanced their manual flying skills. "I used to be a good pilot but now I'm a great pilot." In the words of one, "This particular technical advance really enhances a pilot's stick and rudder skills and his confidence level." Often this sense of control is contrasted to that in more automated aircraft: "So a computer is also watching over us. It's not like with an Airbus, where, famous last words—what's it doing, not working, boom! We are still in charge. We can overrule the computers, but it's helping us."

This sense of control contains a central irony. The HUD certainly does not free pilots from dependence on computers. They are still relying on a system programmed with human skill and judgment. As one test pilot put it, "Flying with the flight path vector equals total trust in software." Yet because the software is only creating symbols, and not actually pushing and pulling the control surfaces (what some people still consider "flying the plane"), the pilots can filter its data, enhance its outputs, and stay more involved.

Indeed, the pilots share their control with the people who make the HUD, so I spent some time talking with them—the technical ghosts in the machines. The entire company of HudView comprises several hundred people, including those in marketing, manufacturing, and quality control, but the core technical team is only about fifteen people. A few, but not all of them, are pilots; most are engineers or programmers.

Mary, for example, came to HudView from the television industry. She was always a little offended at the HUD's monochrome green display, compared to the rich color of television. "But you don't fly your televisions either. So if it crashes, nobody dies and you don't get sued." As it happens, the HUD started out in the green 1980s cathode ray tube color, which has stuck. The FAA restricts the use of colors like yellow and red to warning conditions, and the Day-Glo green seems to be relatively rare in the natural environment.

John plays video games in his spare time, and some of the interfaces he plays influence his work. "They all have interfaces that tell you things," he says of the games, "and that's all very HUD-like. It has to be around the periphery and available, but not block your view of the action. I think that's a strong influence."

Some HudView engineers feel as though they are literally working with the pilots who are landing the plane. One even believes that while sitting in the aircraft's cabin as a passenger, he can tell by the feel of the landing whether the pilot is using the HUD.

Of all the symbols, the guidance cue and flight path vector are the richest and most interesting. Bob is an electrical engineer who's worked on the core aspects of the HUD's control code, proving it in simulators and flight tests, and seeing it through from design, flight tests, and certification. As a private pilot, he is able to bring some personal knowledge of aviation to the job. Today he is one of about three people involved in the core control algorithms.

"It seems kind of intuitive to people that, well, it's just telling me where I'm going," Tom says of the flight path vector, "but there's just a lot more to what it tells you in relationship to all the other symbology on the HUD." With experience, pilots can glean the winds, the aircraft's descent rate, and the trend of its momentum from the flight path vector alone.

The guidance cue, too, is no simple data reporting, but in fact represents a great deal of engineering and human decision, what Bob calls "a negotiation." To him, the guidance cue is "a manufactured product. . . . It's a mix of a bunch of different kinds of data. And there's a certain kind of judgment that goes into it." The cue needs to be accurate, but also to move smoothly without noisy, jerky motion. "If you tried to take the commands that you send to the autopilot and the servos and all that kind of stuff, and gave it to a human, it may be telling you to do things faster than a human can respond. So you have to tune them a little differently." The exact nature of the data filtering is one of HudView's key trade secrets, the recipe for their secret sauce.

Nowhere does this negotiation become more critical than in the flare, the last maneuver the aircraft executes before landing, when the nose pitches up and the plane gradually stops flying until its main wheels touch down on the runway. Commercial pilots take great pride in the smoothness of their landing flares, and there is not one best way to do it. Some pilots like a deep, brisk flare, whereas others like a slower, steady round-out to touchdown.

Landing conditions can vary widely in wind speed and direction, the slope of the runway, snow and rain conditions, even the altitude of the airport. Pilots use their skill and judgment to adjust the flare for the smoothest possible touchdown.

But not the HUD. For Bob, and HudView, the priority is uniformity over grace or perfection: "The system has to do this the same way every time. . . . We can't look at it and say, OK, that's John Doe, it'll be this flare for you, and some other [flare] . . . for someone else."

The HUD has specific parameters it needs to optimize, like accuracy and repeatability, in order to be certified by the FAA. "It's like threading a needle," Bob says. At high-altitude airports in the mountains, for example, the air is thin so the airplane is coming in relatively fast. "You have to shorten the flare time, and the guy has to be right on it."

Some pilots feel the HUD makes their landings more consistent. "Greasing it in [a perfectly smooth touchdown] is great," Bob says, "but then you'll know that right away the trade-off for that is your touchdown footprint is going to get big because of the inconsistency." In bad weather the trade-off is even more pronounced: "Our CAT III flare guidance sits you down firmly . . . so that it makes hitting the point on the runway more consistent and trades that for a very, very soft landing." More trade-offs: "It's the consistency against the soft."

To enhance the flare, pilots may choose not to slavishly follow the HUD's guidance cue. In this case the pilot "might just pull back a little, raise that [nose] just a hair. He knows that if he does that, he's going to get this real nice landing." Rather than seeing such enhancement as a work-around to his finely crafted and coded guidance cues, Bob finds this tweaking part of the point of the HUD—the person is able to alter the recommended flight path according to human priorities, desires, and skills in a particular situation. "I can easily sweeten up the guidance," Bob imagines pilots saying to themselves, "by bringing the flight path a little bit closer to the rise"—meaning a smoother landing. This user-aided landing might prove to be the most significant aspect of the HUD's engineering.

In 2009 the Flight Safety Foundation, an independent, nonprofit organization, did a study of the HUD's potential safety. They looked at nearly a thousand commercial and corporate accidents over a twelve-year period (1995–2007) and tried to determine what effect a HUD, if used, might have had. They concluded that the modern, large-view conformal type HUDs might have prevented 38 percent of the accidents under study and nearly 70 percent of the takeoff and landing accidents. The HUD contributed to safety in several ways, but at the top of the list was the flight path vector, followed by the acceleration and speed error tapes, flare guidance, and the guidance cue.

It is likewise not difficult to look at recent high-profile accidents and see how a HUD might have prevented them. In the 2009 Colgan Air crash in Buffalo, where pilots allowed the aircraft to slow dangerously into a deadly stall; using a HUD the pilots might have seen the decay of their airspeed and the aircraft's energy state soon enough to fix the problem with greater margin for error. In a 2009 Turkish Airlines crash in Amsterdam, the crew might have noticed sooner that the autoland system was not operating properly due to a faulty sensor and, as with the Colgan crash, might have noticed the decay of the aircraft's energy state. In a 2013 UPS Airlines crash in Birmingham, Alabama, in which the pilots conducted a "non-precision" night approach into a "black hole" and struck a hillside, they might have seen their way more clearly to the runway using a HUD.

As described earlier, in the summer of 2013, an Asiana Airlines Boeing 777 approached to land in San Francisco. That day the pilots were working their most basic task: land at a modern airport in clear weather. The Asiana crew had no heads-up display, and the standard radio glide slope from the airport was out of service, although its visual equivalent, a system of fixed runway lights that show the aircraft's position relative to the glide path, was working.

The aircraft never became stable on its final approach—at first too high above the intended glide path and then too low; at first too fast and then too slow. It arrived too early at the runway; the tail struck an embankment and the aircraft spun around and caught fire, killing three, injuring dozens, and destroying the aircraft in the first incidence of fatalities in four and a half years of U.S. commercial aviation.

All pilots should be able to land their aircraft visually in clear weather. Yet the Asiana pilots failed to manage their flight path and their airspeed, something every beginner pilot is trained to do. The Asiana pilot flying said he was stressed about landing without the aid of the radio glide slope. What's more, the airline's policies encouraged the pilots to use "as much automation as possible," which many interpreted to mean using autoland. Even so, the pilots did not adequately understand the logic by which the automatic throttles operated. In training, the correct operation of the auto throttle had been dismissed by an instructor as an annoying anomaly. We can imagine the instructor tragically saying, "It does that sometimes."

The Asiana pilots hesitated to fly manually for fear of retribution if something went wrong. In 2012, only 17 percent of Asiana landings used autoland, but in most of the 77 percent of manual landings the pilot took over manual control from the autopilot only below 1,000 feet above the runway, when much of the work had already been accomplished by the machine. "Without greater opportunity for pilots to manually fly the airplane," the accident report concluded, "their airplane handling skills degrade."

Might a HUD have alerted the Asiana pilots to the decaying energy state of the airplane with enough time to remedy the problem? Might the hands-on nature of HUD flying have prevented the decay of their manual flight skills?

More controversially, even in the en route phase of flight, might a HUD actually have helped the Air France 447 pilots correct the attitude of their airplane and prevented the fatal stall?

My goal here is not to evaluate HUDs, nor to promote their benefits; they are clearly not a panacea for cockpit automation problems. No HUD would have prevented the Asiana pilots' confusion about the auto throttles, for example. Whether the HUD improves safety in a statistical sense will become apparent over time.

Rather, my point is that HUDs represent a new approach to the problem: an innovation which, while certainly "high tech," is an innovation in the humans' role in the system. Rather than sitting back and monitoring, the pilots are actively involved. Again, sometimes *more automated* is actually a less sophisticated solution. Sometimes it requires newer, more innovative technology to incorporate the human in deep ways. As you sit in an airliner descending toward the ground on a cloudy day, how deeply would you like your pilot to be involved?

The example of HUDs suggests that as we look for solutions to automation problems that arise in aviation and other domains, we should seek innovations in how we combine people and machines, rather than simply adding additional equipment and software. Some of these innovations have been called "information automation," which provides data in new forms to human pilots, as opposed to "control automation," which actually flies the aircraft for them.

As a final example, "synthetic vision" continues the trend of structuring pilots' vision by giving them computer-generated representations of terrain and airports. When flying at night or through clouds on an approach, synthetic vision shows a virtual landscape as well as a flight path vector. For smaller aircraft without HUDs (including my own Beech Bonanza), the display appears behind the glass-cockpit instruments; the flight path vector can now be superimposed on the synthetic terrain, allowing the pilot to "put the thing on the thing"—put the vector on the image of the runway—and fly it in. You'll land right where the vector says you will.

The synthetic view can also incorporate quantitative indicators like compass headings and warnings of obstacles, as well as providing traffic information. Pilots like synthetic vision because it allows them to return to visual flying, in any type of weather: "They learned how to fly visually," Bob, the engineer at HudView, observes, "and now they're going back to 'Just follow this symbol over this other symbol and I'm done.'" Every day is good weather in the virtual world.

Still, synthetic vision highlights the concern with all information automation. True, it places pilots more into the control loop and allows them to manipulate their aircraft directly. True, it can be easily either ignored or enhanced if the pilot is so inclined. But information automation is still heavily dependent on software, created by human beings and human processes and subject to the same phenomena as any other human activity.

Synthetic vision relies heavily on its database, providing the numbers

Synthetic vision display of final approach in the author's general aviation aircraft. Note the circular flight path vector and the outline of the runway. The aircraft will land at the point where these two icons overlap.

(PHOTOGRAPH BY NAT SIMS)

to model the terrain, thus raising the questions: How accurate is the model? How current is it? The database models the world at some moment in the past, and might not reflect airport construction, towers and cranes nearby, even equipment failures. Moreover, the synthetic display only portrays a smoothed, ideal, platonic world, and does not include messy interruptions like deer or trucks blocking the runway. When synthetic views are overlaid on HUDs, the smooth contours of the virtual earth can look like fuzzy, distracting clouds overlaid on the real environment. Still, the compelling graphical imagery can attract pilots to a credulous dependence on the data.

With these caveats in mind, we can now consider the future. Will we, should we, be flying on completely unpiloted airliners? Technologies exist today that can taxi an aircraft to a runway, take off, fly to a destination, and land with no human pilot. But they have been applied mostly to fully unmanned aircraft, where a crash does not equal the loss of human life. Put a hundred or so people on board, and flying without a pilot makes us uneasy.

Could engineers really imagine every possible contingency or emergency that might happen, and program it into the software? Perhaps not, but what if the aircraft and its software can "learn" from every anomaly that might arise (including the automation failures), not just in its own life but in the life of the entire fleet, and incorporate those lessons into its decision making? In an era of fast computers and cheap storage, every aircraft can have a database of every prior flight in that aircraft, in that aircraft type, by that airline, at that airport, and so on.

Researchers (and airlines) are hard at work on extracting knowledge from such big data sets. Such statistical techniques work well for identifying possible fraudulent transactions for a credit card company, or for selecting the ads to display on your next visit to a Web site. But can such predictions be so accurate, so precise every one of tens of millions of times, that we will trust our lives to them, much as we do to those clean-cut, professional but fallible and possibly tired men and women who greet us as we board our flight? If they can, how will we know? We are only just beginning to learn how to certify such algorithms for safety.

The story of heads-up displays is but one among many of how the role of the pilot is changing in response to robotics and automation. These become especially relevant as piloted aircraft are being pressed to share airspace with unmanned systems. The FAA is under political and commercial pressure to open up its highly regulated airspace to unmanned aerial vehicles (UAVs) for agriculture, real estate, even package delivery and moviemaking.

A mix of technology, procedures and regulations will eventually work the problem through. But our experiences from extreme environments, and from the history of aviation, tell us that unmanned aircraft will not mimic manned aircraft, but rather that the two will change together. Much as the submersible *Alvin* incorporates computers and software developed for autonomous robots, manned aircraft are already changing with HUDs, synthetic vision, and other computer aids as they did in earlier generations with autoland, autopilots, and even round-dial instruments.

In the airspace, we're likely to see convergence. Flying an airliner is already like flying a drone that you sit in, and, as we'll explore in Chapter 6, is likely to become more so. Flying a drone should be like flying an airliner, even if the human input comes ahead of time. But when people are on board, the pilots' physical presence serves a purpose, if a social one as much as a technical one. What happens when we move the pilots' bodies to another place altogether, is the subject of the next chapter.

CHAPTER 4

War

WALK OUT OF THE DESERT HEAT OF THE AMERICAN WEST AND into the darkened trailer that houses the Predator control room and you are immediately transported into a remote war zone. Screens glow and equipment fans whir. Like an airline flight deck, the room has two "hot seats": one for the pilot, who commands the mission, and one for the sensor operator, who does the looking. The pilot monitors nearby friends and enemies and communicates outside the control room via radios, chat rooms, telephones, and other devices. The sensor operator commands the camera on Predator, searching for and locking onto targets. A third person, the mission coordinator, sits behind the pilot and sensor operator, keeping in close touch with troops on the ground in the combat zone, intelligence analysts, and others in the chain of command.

To start the mission, a launch and recovery crew located in a theater of war far across the globe prepares the unmanned aircraft, about the size of a commuter plane, and sends it off on its way. Then they hand off control through the global network to the remote crews. While the

Ground control station of a Predator aircraft. The pilot sits at front left, the sensor operator to his right. Note the large number of LCD screens, many of them added by the crews in a burst of user innovation. Note also the whiteboards to the right and left of the crews, used to coordinate data and actions.

(IMAGE BY BRIAN W. JONES, USED BY PERMISSION)

The Predator Remotely Piloted Aircraft (RPA). Note the sensor ball hanging below the nose on the left, and the Hellfire missile slung below the wing.

(U.S. AIR FORCE PHOTO/LT. COL. LESLIE PRATT)

aircraft wings its way to the battlefield on autopilot, the operators prepare for the mission. Predator is relatively slow, so the journey could
take several hours. Much of the remote crews' work during this time
seems like our everyday office work—signing in, logging on, setting up
screens and menus.

Despite the mundane nature of the numerous systems' setups and
checks, they serve to connect the crew with their aircraft and the faraway
situation they are preparing to observe and engage. As they configure
their workplace and tools, customize their displays, and set their personal preferences, they develop a sense of immersion and presence.
"Who is on the Internet chat rooms?" they ask. "What is going on in the
field? What is going on on the network?"

Frequently the work shifts do not coincide with the start or end of
a mission, so a new crew observing the last few minutes of the prior
shift gains awareness; during particularly intense periods, the crew
shift might be delayed. "It's a very big paradigm shift," said one F-16
fighter pilot turned Predator pilot, used to climbing into a jet on the
ground to start a sortie. "I will walk into the cockpit and there is already
a crew doing the mission."

Predator remains prone to computer crashes and lockups during
critical periods. Crews learn through difficult experience not to press
certain key combinations, not to issue commands too quickly, not to
confuse this button with the one next to it.

It takes several steps to do simple things. It takes more than twenty
keystrokes, for example, simply to turn on the aircraft's autopilot. "The
conjecture among us pilots," observes one operator, "is [that] the engineers thought we were too stupid, and we would be idiots and be hitting
buttons all the time and doing dumb things, so they tried to put a two-
step process into everything that we do."

Manuals are lengthy and unclear. Some important features are hidden in the system's code and documented nowhere, passed down through word of mouth between the operators in a kind of oral tradition of stories of work-arounds to make the system perform. No small part of the operators' skill is simply making the system do things it was not designed to do.

What constitutes "flying" for Predator pilots? From the handover immediately after takeoff, much of the mission is conducted on autopilot; the pilot monitors the vehicle as it automatically follows a series of GPS-derived waypoints. Even through this remote connection, however, the pilot still retains the ability to hand fly the vehicle using the joystick. Yet because of the absence of physical cues—g-forces, feelings of turning, even the engine sounds, vibrations, and smells—hand flying presents a challenge.

The difficulty of remote flying is exacerbated by Predator's design. One of the most basic controls in an aircraft, from a training Cessna to a fighter plane, is the ability to "trim" the aircraft—setting a point to establish the aircraft at a particular pitch angle and airspeed. A properly trimmed aircraft requires just a light touch on the controls by the pilot. But Predator has an unnatural, cumbersome, multistep screen, stick, and button process to trim the controls and stabilize the vehicle.

Even worse is the location of buttons on the control stick: on U.S. Air Force aircraft the button to drop bombs is located, by convention, at the top and left of the control stick. On Predator, pushing the button in that location turns off the aircraft's stability augmentation system, which can easily send the aircraft spinning out of control.

Hand flying remotely is also complicated by the 1.8-second time delay to issue commands to the vehicle and receive a response. One might logically assume that the time delay derived from the necessity for sending commands halfway around the world via satellites in space. Yet only about half a second of the control delay is due to the speed of light. The remainder of the time delay occurs in video compressors, routers, and all the other equipment that processes the data. The communications system was designed to optimize image quality, not speed of response.

Pilots can command the aircraft to fly to and hold specific headings, and can even use the stick to command changes in the autopilot (similar to the fly-by-wire techniques that control many of today's airliners). Later versions have features like point-and-click loiter, which enables the pilot to identify a point, spread a circle from it, and have the aircraft automatically fly around the point—helpful in quickly setting up observational sorties.

Still, the Predator autopilot can only command the aircraft to bank fourteen degrees to each side, limiting the rate of turns. Pilots sometimes take manual control in order to command twenty or thirty degrees of bank for more extreme maneuvers. The trouble with these bank angles, and the reason the autopilot will not command them, has less to do with the aircraft itself than with the servos that point the onboard satellite antenna up toward space. It will lose its signal lock at higher bank angles (the way the satellite TV signal in a modern airliner will drop out during a turn). If pilots command too great an angle, they have to be aware they could knock the bird offline for a few seconds and lose all image feedback and control authority until the autopilot rights itself and again acquires lock—a risky maneuver.

Predator pilots, in public presentations and memoirs, tend to emphasize the continuing need for their hand flying skills in specific situations such as avoiding weather, obstacles, or other aircraft. Yet others within the air force have called these control sticks simply "morale sticks" and feel that flying Predator manually through the satellite link is "stupid." In the words of former air force chief of staff General Michael Ryan, "We shouldn't have pilots stick-and-ruddering UAVs." Rather, the argument goes, the pilots should be using the autopilot and way-point following, controlling the vehicle at a higher level of abstraction.

By contrast, generating the imagery and videos is "a matter of life and death," but one squarely in the domain of the sensor operators and not the pilots. They find themselves employing new skills unrelated to aviation. They calibrate the cameras and the infrared detectors to the immediate temperature, environment, and time of day. They work with the pilots to put the aircraft in the best position to see things, and then adjust focus, gain, and camera level to create "a statistically significant scene"—one that can yield meaningful information. Sensor operators call this work "growing" strong "video tracks"; they treat the tracks "like living things that need periodic attention," and take pride in the quality of their video.

The imaging system has some automated ability to make these adjustments, but often the human operators can do better than the computers because of their knowledge of the human context. For example, they can tell the difference between a running engine and a cold one, or discern people from livestock, or adjust for the time of day and the desert heat. Sensor operators can coordinate all these data and imagery with information from the radios and chat rooms. Bandwidth over the satellite link is a limited resource, and the operators trade off certain kinds of data for others, putting priority onto the video feed for the best image quality during critical periods.

Floating above the ground, looking down on their quarry from above, Predator sensor operators grow "to feel as if they were the sensor ball itself," referring to the little dome of cameras, lasers, and servos that hangs below the nose of the aircraft. Sitting in their trailer, pilots and sensor operators crane their necks and move their bodies as they seek to look above or around an object on the screen. They will sometimes whisper to one another during tense moments, even though their loudest shouts could not be heard far away on the ground. Instructors believe that such feelings of presence allow the sensor operators to focus on a scene, become more curious about what they see, quickly detect movement, and rapidly react to anomalies.

Fighter pilots who actually fly above a battlefield acknowledge that they do not feel as deeply present as the sensor operators. Because of a combination of factors, ranging from the size of the screen to the presence of other people, the sensor balls on their jets do not provide the same sense of "being there."

War, as structured state-sponsored fighting and dying, has given rise to venerated professions to manage its risks and bound its ethical dilemmas. The nature of those professions, and the meaning of their experiences, changes when the human body is no longer present in the fight, fueling a public debate. This chapter, as a biography of the drone, explains how Predator came to be, and how it operates today, showing us remote operators deeply involved in the events they witness and affect. As with remote explorers undersea, their experience stems as much from social relationships as from the machinery itself. As in the airlines, the identities of the aircraft and those of its pilots have evolved together, each struggling for stability and recognition.

The U.S. Air Force's Predator and Reaper remotely piloted aircraft (RPAs), have become the public face of "drone warfare." Predator has gone far beyond an experimental stage and operated more than a million combat flight hours by 2010, and a million more by 2013. The air force has invested heavily in Predator, purchasing hundreds of aircraft to operate more than sixty combat air patrols (CAPs) (though losing more than seventy Predators to crashes). In 2012, the air force put more RPA operators through initial qualification training than pilots for fighters and bombers combined.

Each CAP, capable of keeping a continuous, twenty-four-hour presence over a target, is composed of four aircraft. Far from being "unmanned," a CAP requires more than a hundred and fifty people to operate it. As of 2009, Predator operation required nearly 4 percent of total air force personnel, and more than 9 percent of its inventory of pilots. Predator A production ended in 2011, but Predator B—a more heavily armed version with twice the speed and range and ten times the payload, also known as Reaper—continues rolling off the production lines.

Predator's distinctive shape, with its bulbous nose where a cockpit would normally be, its thin, long wings, and its downward-canted tail hiding a pusher prop, has become an icon of a conflicted robotic era. It can loiter, nearly silent and invisible, over a distant battlefield for long periods, watch with intense focus through infrared cameras, and then unleash precisely targeted small bombs or missiles. This technology, depending on whom you ask, has become either the cutting edge of future warfare or the symbol of all that is wrong with American technological power, harbinger of an era of autonomous robotic surveillance. Whichever side of the debate you're on, Predator has been a focal point.

Yet despite the profusion of press coverage, controversy, and public conversation, we know relatively little about what Predator operators actually do, and how the technology affects their experience of war. While often compared to manned fighters and bombers, Predator is not just a wannabe airplane, and its pilots are not wannabe pilots. Through an unusual and circuitous evolution, Predator has evolved into a new kind of aircraft, doing a new task, driven by new kinds of specialists. Much of that evolution has been led by Predator's users, whether the people operating the aircraft from the ground, the people fighting in the theater whom it supports, or commanders who deploy it in warfare.

Through these innovations Predator pilots feel deeply present on the battlefield, albeit in ways mediated by technology. "The Nintendo mentality is a detached mentality," one operator observed, responding to criticisms of "video-game" warfare that is "deceptively antiseptic." "This stuff is real. I'm taking real lives. I'm shooting real weapons. And I have to be really responsible for my actions."

Predator's task is not that of the lone aircraft, the lone pilot, or the autonomous drone. Rather, it is as part of a global system, with many linked users and consumers operating from vastly different locations. For some, it represents a revolution in warfare, for others simply a new military tool in the arsenal, for still others the crossing of a troubling ethical line governing remote killing. But what Predator is *not* is an autonomous system, making decisions on its own and operating free of human control.

Unmanned aircraft have been part of the U.S. military for decades. Each era of aviation had its visionaries who saw them as an inevitable, progressive evolution. Yet until the turn of the twenty-first century, UAVs remained, in the words of historian Thomas Ehrhard, "peripheral systems locked in a seemingly endless, inglorious loop." To understand Predator, we must understand that inglorious loop and what broke out of it.

Unmanned aircraft go back to the dawn of aviation, and the U.S. military has experimented with them since the 1920s and '30s. In 1936

Lieutenant Commander Delmar Fahrney, running an obscure office within the Army Air Force, coined the term "drone" to distinguish American vehicles from a similar British aircraft dubbed "Queen Bee." During World War II, numerous experiments in unpiloted aircraft and missiles produced everything from brilliant inventions to ignominious failures.

The Nazi-developed V-1 "buzz bomb" entered combat right after D-day in June 1944; of the nearly ten thousand launched at London in the ensuing months, about 23 percent found their targets, killing more than six thousand people. The V-1s had little "intelligence" in today's terms, but they did have basic autopilots and feedback loops that kept them flying in very straight lines and at level altitudes. Ironically, these automatons were *too precise* at piloting—their arrow-straight trajectories made them easy targets for new radars and electronic computers. Nonetheless, they earned the term "robot bomb."

Undoubtedly, being on the receiving end of a V-1 made one feel the target of an evil, focused intelligence. The feeling was accurate, though the intelligence was not that of the automata but of the Nazi engineers who built them, and their leaders who targeted civilian populations. The V-1's larger and more expensive sibling, the V-2 ballistic missile, was also a robot weapon, setting up a technological rivalry between rockets and pilotless aircraft that would shape the technology for decades to come.

After the war, U.S. militaries devoted themselves to the development of guided missiles. The "New Horizons" report of 1945, famous within air force circles for laying out the technological road map of the next fifty years, had an entire section on "unmanned devices," and by the following year the U.S. military had nearly fifty guided missile projects in the works. These included the now-forgotten Regulus, a rocket-launched, jet-powered, radio-guided cruise missile with a nuclear warhead that was about same size and weight as a fighter plane. Others included Matador and Navaho—expensive failures. Efforts to convert manned bombers into unmanned platforms also produced little success.

The 1950s term "guided missile" reminds us that whatever trend in military robotics led to Predator, we have long been in a world where some form of automatic guidance is built into weapons. Guidance systems made missiles as much robot automata as any other machine of their age. While confined to predetermined trajectories, they took a great deal of feedback from their environments and incorporated it into control loops, including taking star sightings during their brief periods of spaceflight to precisely align their guidance systems for accuracy.

As we have seen elsewhere, less sophisticated know-how from the predigital era made unmanned aircraft more automated and more autonomous than today's UAVs. Predator is *more* under human control than these earlier systems.

Unmanned aircraft were caught in a technological middle ground between manned aircraft and guided missiles (especially the nuclear intercontinental ballistic missiles). Both these rivals were platforms with clear missions, compelling public imagery, and well-defined advocates and constituencies. The advent of satellites for surveillance provided yet another source of competition.

When the Soviets shot down Francis Gary Powers's U-2 spy plane in 1960, the American response was not to build unmanned aircraft. Rather, the incident led to faster, higher manned aircraft (the exotic Mach 3 SR-71 spy plane) and more emphasis on satellite reconnaissance, favoring these more mature technologies. A supersonic UAV launched from an SR-71 was developed at great expense in the 1960s, flew operationally four times, and was canceled due to failures, high costs, and more effective imagery from satellites. Even today's satellites are, of course, robotic vehicles employing a mix of automation and human control from the ground.

In the late twentieth century, unmanned aircraft did find a niche as returnable target drones used as gunnery targets for training. These were either small, custom vehicles or manned aircraft with the pilots replaced by computers and guidance systems. The Ryan Firebee drone, for example (made by the same company that made Charles Lindbergh's *Spirit of St. Louis*) could execute a preplanned route and return to a recovery point and land by parachute. Over four thousand Firebee drones flew prior to 1971, and the extensive experience spurred more operational uses.

Firebee-derived vehicles saw combat in Vietnam, in a reconnaissance version known as Lightning Bug, launched from C-130 aircraft. Most were decoys or electronic jammers; some recorded data, especially on electronic emissions of antiaircraft radars. Most were flown on preprogrammed flight paths, though some could be remotely controlled by a pilot aboard the launch aircraft and return their data in real time. From 1964 to 1974, the United States flew more than 3,000 missions against China, Vietnam, and North Korea with the loss of 578 aircraft. A few even flew in Iraq in 2003. Despite this record, these vehicles remained inflexible, vulnerable to loss, and dependent on specialized support aircraft for launch and control.

One explanation for the halting adoption of unmanned aircraft blames pilots who resisted them out of wariness of losing their jobs—and their glory—to clockwork automata. But scholars have found little evidence of this "white scarf syndrome"—the traditional defense from pilots unwilling to give up their heroic image. In fact, in the pilot-dominated air force, where resistance could be expected to be stiffest, the record shows repeated, active attempts by pilots to develop and deploy unmanned aircraft. Still, pilots do love their aircraft, and cling to them, while unmanned aviation lacked passionate advocates, with "only the obscure novelty of a mechanical feat and the promise of saving lives to propel it."

But the drones also came up against other barriers. Routinely sold versus manned aircraft on economic grounds, unmanned aircraft had trouble delivering comparable capability at reduced costs. Their missions were often poorly defined. Treaty limitations, especially the Intermediate-Range Nuclear Forces (INF) Treaty, designed to limit nuclear weapons, did not recognize subtle technical differences, and defined cruise missiles in a broad way that included unmanned aircraft, which limited their legal uses.

Technical limitations played a role as well. Until the advent of GPS in the 1980s, the difficulty of navigating unmanned aircraft proved almost insurmountable; missiles and aircraft relied on inertial measurement systems too heavy for small aircraft. Data links were heavy, limited, and unreliable, as was onboard computing.

Perhaps most important, unmanned aircraft were always competing with mature, better-funded systems—especially aircraft and satellites—with established communities of support and personnel behind them. Nearly every technological improvement that benefitted unmanned aircraft—computing, navigation, electronically controlled weapons—also benefitted the better-established, more mature, and more familiar manned systems.

Due to the twin challenges of technological immaturity and competition from other technologies, then, unmanned vehicles were always playing catch-up.

———

It was against this backdrop of expensive failures, technological limitations, and niche applications that Predator struggled to be born. In Ehrhard's words, "A more serendipitous weapon system program is hard to imagine." Far from a disruptive breakthrough or an obvious, linear evolution, Predator evolved in fits and starts. Today, the aircraft's peculiar biography profoundly shapes the identity and experience of its operators.

The project began in the 1980s as a special program by the Defense Advanced Research Projects Agency (DARPA) to build a long-endurance reconnaissance vehicle that could also serve as a cruise missile. The prototype, called Amber, was built by a small American company called Leading Systems Incorporated (LSI), under the direction of its founder, charismatic Israeli designer Abraham Karem. Amber first flew in the late 1980s, powered by a small piston engine designed for snowmobiles and recreational aircraft. In 1988 it stayed aloft for more than thirty-eight hours, a record for unmanned aircraft.

Faced with inefficiency, duplication, and failure in unmanned programs, in 1987 Congress ordered a consolidation. DARPA transferred the Amber program to the navy, which promptly canceled it. LSI went bankrupt, and in 1991 the company was purchased by General Atomics, a contractor with little experience in aviation that also hired Karem and most of his engineering team. Under Karem's direction, General Atomics continued working on an LSI vehicle called GNAT-750, derived from Amber, for export.

Today, Karem and his team are celebrated in the press for inventing the drone. Indeed, they created an airframe that was light, long-endurance, and more robust than its competitors. "I just wanted UAVs to perform to the same standards of safety, reliability, and performance a manned aircraft [would]," Karem said in a recent interview. But Karem and his team were aircraft designers; they did not envision the vehicle as part of a larger system of interacting people and machines. The larger significance of Predator, while requiring a dependable airplane, stems more from its integration into global networks than its design as an aircraft.

General Atomics sold a few GNAT-750s to Turkey in 1993. When the conflict in the Balkans began escalating, the CIA came looking for long-endurance surveillance of Serbian targets. Soon GNAT-750s were flying over Bosnia. Their pilots were a mix from all the services but predominantly army helicopter pilots. They controlled the craft from Albania, but could relay video back to the United States via satellite. The aircraft couldn't fly in the Balkan winters, so they quietly packed up and went home to await better weather. The mission ended after persistent problems with icing, and after the loss of two aircraft to ground fire and accidents.

The U.S. military had a dearth of assets for "tactical reconnaissance"—close-in views of dynamic ground forces—so the GNATs attracted the attention of General John Jumper, commander of air forces in Europe. The first Iraq war in 1991, when U.S. forces had difficulty locating and destroying Scud missile launchers, highlighted the need to track and follow mobile targets. With the GNATs, Jumper saw real-time video feeds that monitored hostile forces' movement of weapons, aided in search and rescue, and oversaw humanitarian operations (one even covered Pope John Paul II's route on his visit to Bosnia in 1997). Despite failures and crashes, GNAT came to be known for its constant, unblinking stare, what became known as "persistent surveillance."

General Atomics enlarged and upgraded the GNAT-750 to add a satellite link so it could be operated from great distances. In 1994 they also upgraded the name from the diminutive GNAT to the ominous "Predator"—an ironic name given that the vehicle was unarmed. Whereas the original vehicle had to be within line of sight to the ground station (about 120 miles), a new satellite link extended the range to anywhere within the same ground "footprint" of the satellite transponder (about a 400-mile radius).

Predator acquired its iconic bulbous nose to house the satellite antenna; it also acquired the characteristic sensor ball cameras and other sensors hanging off the nose in a round housing. This vehicle also benefitted from the new ubiquity of GPS, which finally allowed precise positioning, the lack of which had hampered earlier unmanned craft (indeed, Predator was the first operational unmanned aircraft to use GPS). Moreover, unlike its predecessors, Predator could take off and land on traditional runways, with no requirement for specialized support aircraft.

Predator began proving itself with positive results in tests and exercises and saw its first combat deployment from Hungary, flying over Kosovo, in 1995. It took eight hours of flying time for the slow craft simply to get to the target area. But as near real-time video downlinked to operators in Hungary began crossing the Atlantic and streaming into Pentagon offices, Predator crew members were surprised to begin receiving phone calls from generals and admirals, newly addicted to the video feeds. The crews began to refer to the phenomenon as "Predator porn."

Attracted by this new ability to peer down on the enemy, in 1996 Secretary of Defense William Perry formally gave the Predator program to the air force, which began the slow, tedious work of turning the development program into a formal acquisition. Faced with a lack of procedures and documentation, in 1998 the air force turned the task over to an internal group called Big Safari.

Ordinarily, this secretive outfit quietly took advanced, commercially available components and combined them into new systems for unique missions, usually building just a small number of any given platform. Big Safari was known for innovation, not bureaucratic skill at producing documents.

It tried to normalize the program, often in conflict with the General Atomics engineers, who considered themselves disruptive innovators not answerable to military bureaucracy. The company had developed Predator largely with internal funds, so the engineers were reluctant to turn over data to the government for standardization. While some of the engineers were pilots, they tended to be from the general aviation world, flying small aircraft on weekends, rather than from the military, where aircraft needed close cooperation with larger systems.

General Atomics engineers constantly clashed with air force pilots and program managers who emphasized documentation, predictability, and uniformity. "It took [General Atomics] a while to grasp the fact," Big Safari leader William Grimes wrote, "that they could not go to the field and make unilateral changes without telling anyone."

Finally, in 2000 a "baseline" Predator appeared, supposedly ready for incorporation into air force operations for persistent surveillance. It included a more powerful engine and radios that would communicate with friendly aircraft in hostile airspace.

Karem and his team's numerous clever shortcuts to make Amber, GNAT, and Predator inexpensive, simple, and light had the unintended consequence of making the aircraft difficult to fly. The long, slender wings made it tricky to land in gusts and sensitive to icing. The aircraft's systems could not sense the amount of fuel on board, but only derived the information by monitoring inaccurate fuel-flow meters. No switch sensed whether the vehicle was actually on the ground or in the air, leading to confusion on aborted takeoffs. Numerous other shortcuts plagued the system for years.

These troubles stemmed less from sloppy engineering than from the evolving identity of the machine. Karem and his engineers were interested primarily in designing an autonomous vehicle. As research and development engineers excited by pushing the envelope, they focused on light weight, endurance, and reliability, not on adapting to the users or larger networks.

Because they emphasized autonomy, they downplayed the ground control station. Why build a fancy interface, they reasoned, for an aircraft that was supposed to work on its own? In fact, they believed that the less input given to the human, the more reliable the vehicle could be. Yet again, it seemed a more elegant, tractable problem to build an autonomous system to work on its own than one that had to work within human systems.

The ground station, which survived nearly unaltered from Amber to Predator, suffered from this prototype nature. The user interface for Predator's operators was designed by engineers and for engineers. Most of the critical flight data, for example, appeared on the screen purely as numbers rather than the dials or tapes that pilots were used to. While the pilot had a traditional joystick, the primary tool of interaction was a keyboard and mouse or trackball, manipulating a complex menu system up to five layers deep. Hit the wrong key at the wrong moment, and you could easily lock up the software, or kill the engine.

Big Safari developed the prototype into a fieldable system quickly and relatively cheaply, but it took years for Predator to be properly integrated into the air force's acquisitions and logistics systems. The clunky flight manual for Predator was more than 1,500 pages long, and still left out critical information. The standard checklist, designed to sit on a pilot's knee, comprised 198 pages of index cards.

To fly the Predator, the air force needed a source of pilots. In 1995 the service reactivated a dormant group, the 11th Reconnaissance Squadron (which had operated drones in Vietnam and into the 1970s), to fly the Predator. It was led by a rated pilot—"rated" being the air force lingo for qualified aircrew, whether pilots or other specialties like navigators. Ironically, the 11th was based at Nellis Air Force Base in Nevada, home of the air force's equivalent of the navy's famous "Top Gun" school.

By contrast to that elite, the 11th consisted mostly of tanker and transport pilots "voluntold" to fly Predator. They felt like outcasts in the desert, isolated from their normal communities. Many had not even been aware of Predator before their assignment. They had no access to manuals, specifications, or operating instructions from General Atomics. They found themselves being trained by army helicopter pilots who had learned to fly the system largely by word of mouth.

The lack of formal documentation and procedures gave the whole affair a chaotic air that did not improve morale. Above the entrance to their base outside of Las Vegas, crews of the 11th posted a sign: "Leper Colony."

Unarmed tactical surveillance and reconnaissance missions already had low status within the air force, and the unmanned nature of Predator just pushed them further down the ladder of social prestige. Predator still seemed like a toy. Most pilots wanted nothing more than to put in their time and move back to their aircraft. The air force had to entice them to fly Predator by offering plum assignments for their next tour; even so, many pilots hung up their careers and left the air force rather than fly "unmanned." Few who flew Predator ever went back to their manned aircraft.

Staffing problems amounted to tensions over roles and conflicting models of the experience of flying. Were Predator pilots operating in the grand air force tradition of skill, bravery, and command? Or were they systems monitors, staring at computers, pushing buttons, and watching the miles tick away from the isolation of comfy chairs?

The questions were not just philosophical; they had real impact on careers and paychecks. Training and employment standards for the crews were delayed by years because of disagreements over whether they should be designated as "aircrew" or "air vehicle operators." In air force culture, aircrew were pilots, received flight pay and extra pay for combat time, while the "air vehicle operators" did not. The difference between "real" warriors operating on the front lines and the thousands of support staff is fundamental to the military's social structure; Predator undermined that distinction by creating remote warriors.

Pilots of military aircraft experience computerization and technological change as a matter of course. In fact, their jobs have barely been stable since the vaunted figure of the "ace" was invented by PR machinery in World War I. The 1930s and 1940s saw the introduction of automatic pilots, computerized bomb sights, and radio navigation. Jet aircraft in the 1950s brought a host of electronic gizmos like yaw dampers, stability augmentation systems, and even ground control of fighter intercepts, all of which distributed the pilots' tasks among other people and machines. In the 1970s, the F-15's advanced radar and digital computers transformed the pilot's role from that of a stick handler to a cognitive-information processor. It became apparent that if an F-15 pilot even saw an enemy aircraft, he was much too close and in for a very dangerous fight.

In the first Persian Gulf War in 1991, pilots of the F-117 stealth fighter became heroes for their raids on heavily defended Baghdad. Yet during these bombing runs, F-117 pilots held on for automated rides as their computers flew the planes in precise trajectories designed to dodge enemy radars. The pilots monitored their timing and flight paths, locked their eyes on a video screen, and clicked a virtual cursor on an electronic image to aim the targeting laser. Cognitively, these tasks resembled those of the later Predator pilots, with the crucial difference that the F-117 pilots' bodies remained in the airplane.

In the early years, Predator pilots tended to come to the job through one of three routes. Undergraduate pilot training was the most traditional background; some of these graduates had flown other aircraft and others came directly to Predator. Predator pilots might also come from other aircrew jobs such as navigator, electronic warfare officer, or weapons systems officer (by contrast, the army and Marine Corps both use enlisted personnel to fly their versions of Predator, which are slightly more automated but otherwise identical aircraft). Nonrated pilots still had to have civilian commercial pilot ratings from the FAA—primarily so they could work in the continental United States near civilian airspace.

These three routes, of course, produced differing skills and attitudes in the operator's seat. Rated personnel are most likely to have familiarity with aviation culture and procedures, and depending on their level of pilot training, with traditional stick and rudder skills. Yet they also might resent their assignment to an unmanned system and itch to return to a "real" cockpit.

One typical pilot, we'll call him John, had flown the F-16, but had been grounded from the jet due to a medical problem. Flying Predator was his only option to still be a pilot, so he jumped at the opportunity. When asked about the most difficult part of the training, he answered unequivocally (as many commercial airline pilots would): "the human–machine interface . . . it was almost like learning how to fly all over again."

At this stage many Predator pilots were not enthusiastic about their work, and still hoped to return to their aircraft. Flying remotely simply didn't have the personal thrill, or the social prestige, of zooming above the battlefield.

A number of the most successful Predator squadron commanders were nonpilot, rated officers. By switching to unmanned aircraft, they could gain command they would never achieve in traditional roles as nonpilots. The newer, specially trained Predator pilots might lack aviation skills and familiarity, but also might be best adaptable to the unusual RPA experience, as they would not be comparing the experience to anything else.

Unstable crew identities reflected the unstable identity of the aircraft. Predator was conceived as an ISR system—air force speak for "intelligence, surveillance, and reconnaissance." That meant operating like a spy plane, taking aerial photographs. Indeed, arrangements for processing Predator data into forms useful for intelligence was modeled on those for the U-2 spy plane: take pictures and radar imagery and transmit them back to base for analysis.

But Predator had one significant difference from a traditional reconnaissance aircraft. The older platforms focused on what could be seen in still imagery: buildings, construction sites, accumulations of forces. By contrast, Predator shot video, opening up the time domain to what photographs missed: movements and human behavior.

Predator crews found themselves following trucks through crowded streets as unsuspecting drivers led them to weapons caches or safe houses. They found themselves staking out buildings, monitoring the behavior of those who came in and out. They found themselves watching and supporting troops in combat in real time.

Traditional social structures for handling reconnaissance photographs could not keep up with the need for real-time interpretation and storage for later analysis. One early Predator pilot was shocked to find intelligence analysts printing out frames of video as 8x10 glossies—the standard currency of aerial reconnaissance. To him the essence of the data was the motion.

In early deployments Predator was not certified by the Defense Department to directly connect to classified networks. So operators set up a separate "mission operations cell"—really just a tent sitting outside the Predator trailer—that took the video feed, digitized it, and inserted it into secure networks for passage to intelligence analysts (with significant loss of video quality).

In the Balkans deployments, other than this one video feed, the entire system was separated from global networks by an "air gap." Target coordinates passed into the control van via floppy disk, or by the "sneaker net" of people walking in and reading numbers. On the receiving end, those watching the video saw a disconnected feed, with little ability for feedback and direction to the vehicle itself and its sensors. Similarly, pilots and sensor operators had little awareness of what they were doing within a larger picture.

In order to communicate with front-line aircraft, in this early setup Predator crews had to speak through a cumbersome series of voice relays, passing coordinates or verbally describing targets. General Jumper derisively called it a "dialogue of the deaf."

Predator operators would have to overcome these and other vestiges of earlier, different identities for the aircraft and force the vehicle to become something new. The enabling innovations had less to do with flying the aircraft itself and more to do with its integration into a global military system.

The first of these were hardware additions to the airframe. General Jumper returned from the Balkans to head the air force's Air Combat Command. Impressed by the potential of what he had seen, he pushed for improvements.

The original sensor ball had three cameras that could find and examine a target (closely enough "to look down blouses," in the words of one operator) but it took a great deal of effort from the crew to hold the image still. With some machination, crews could extract map coordinates from the imagery, but the numbers contained up to a half mile of error. After a few hours of "flying" the crosshairs, sensor operators "got the shakes" and needed a break.

As we saw with *Jason*, one subtle advantage of unmanned vehicles is that they can be modified and expanded quickly and inexpensively as compared to manned systems, because of the reduced needs for human-rated safety certification. On Jumper's orders, beginning in 1999, Big Safari and General Atomics gave Predator a new, cutting-edge sensor ball with significantly improved optics, all in a matter of weeks. The new ball could integrate navigational data from the vehicle to automatically keep the cameras fixed to a set of coordinates on the ground, even as the aircraft moved and turned. Computerized tracking modes could now lock onto an image and keep it centered, and pass accurate map coordinates to other aircraft. Sensor operators could work longer shifts, developing skills for fine-tuning the sensitive and versatile cameras to capture ideal images under varying conditions.

The new sensor ball also had two lasers. One of these could

illuminate a target for the purposes of finding the range to it, which doubled the precision of the map coordinates the system could extract from the imagery. It could also guide a weapon, such as a bomb dropped from another aircraft. A second laser could communicate with friendly troops on the ground by "sparkling" the laser at a target of interest. Ground forces wearing infrared goggles could precisely find the object of Predator's "stare." Further innovations enabled Predator operators to pass video directly to aircraft or troops on the ground with mobile displays, and to speak with them over radio links.

As the century turned, the drone's biography became intertwined with that of Osama bin Laden, and with America's new twenty-first-century wars. Bin Laden's minions had attacked U.S. embassies in Africa in 1998, but after a series of American cruise missile strikes failed to kill him in retaliation, the CIA became interested in tracking his movements for possible future attempts.

Predator seemed an ideal way to covertly watch Bin Laden's training bases in Afghanistan. But the large antennas and imagery analysis units that had accompanied the aircraft in Hungary would attract unacceptable attention in the region. The CIA and the air force thus opted to "split" the Predator control elements.

A small group of CIA, air force, and contractor personnel set up shop in Uzbekistan with the aircraft. They would fly the vehicle locally for takeoff and then send it on its way, then switch Predator's communications channel to a satellite link, enabling control by another ground station at a U.S. Air Force base in Ramstein, Germany. When in control from there, the crews could fly the vehicle and direct its sensor ball, access classified communications networks, and exchange video and voice communications directly with CIA headquarters in Virginia.

Predators began flying over Afghanistan in the summer of 2000.

On their seventh flight, on September 27, they spotted a tall man in white robes surrounded by smaller figures who circled him like guards or acolytes. It was indeed Bin Laden (later analysis of the video showed he had actually first been seen by Predator a month earlier). Despite Bin Laden's self-professed war on the United States, and despite earlier U.S. military attempts to kill him, no action was taken.

Also in 2000, John Jumper had taken over the air force's Air Combat Command. Frustrated by Predator's earlier inability to affect the events it had observed in Kosovo, Jumper ordered another addition that fundamentally changed its character. He ordered Big Safari to begin arming Predator.

Engineers selected the army's Hellfire missiles, originally designed to be shot at tanks from helicopters, because they were small enough to ride under Predator's wings and not interfere too much with its flight characteristics. These 100-pound supersonic missiles (themselves unmanned, human-aimed vehicles) could follow laser designators to a target with great precision and explode a relatively small (10-pound) but lethal warhead. Hellfire missiles were first tested from a Predator in February 2001, and engineering continued throughout the year.

Triggering Predator to kill people in faraway places raised ethical and legal concerns in the pre-9/11 military and CIA. At one point engineers even wired in a red remote switch to the Predator control console, so a designated CIA employee, and not an air force pilot, could pull the trigger with specific legal authority (it was never used). In the summer of 2001 the National Security Council decided not to pursue Bin Laden with an armed Predator.

What's more, mounting such combat missions from Germany without informing the country's leadership was seen as politically and legally troubling. To overcome this limitation, Big Safari's engineers put more miles between the Predator's ground control station and its satellite uplink—in the form of a transatlantic fiber-optic cable. Now, through a "remote split" operation, the operators could work entirely within the United States, clearing up at least some of the politics; Predator's ulti-mate configuration resulted from a mix of technical possibilities and social relationships.

Lasers, missiles, and the remote split were just beginning to coalesce, enabling Predator to emerge from its awkward adolescence, when the United States was attacked on September 11, 2001.

Predators were rushed to the Middle East; by September 15, a CIA landing and recovery element was back in Uzbekistan near the Afghan border. An armed Predator first entered Afghan airspace on September 18. On October 7, the first night of the U.S. air campaign against the Taliban, an armed Predator first fired a missile in combat, its target buildings believed to house Taliban leader Mohammed Omar Mujahid—Mullah Omar. Confusion between the CIA and the air force over how to control the globally distributed weapon, however, bungled the opera-tion, allowing the mullah to escape.

Nonetheless, Predator video was proving addictive all over again; this time its avid viewers included President George W. Bush. Soon battles in Afghanistan, and then in Iraq, were proving Predator's ability to support American troops on the ground. From this point, continued evolution of the system had to occur not as an experimental project but within a military at war, and within a Predator community stressed to its limits by ever-increasing operational tempos and demand for imagery.

Lasers, missiles, and the remote split transformed Predator from a roving eyeball into a global weapon, with profound effects for its crews. No longer were they simply operators, watching the vehicle mindlessly fly its preprogrammed trajectory. They could now kill people and break things, making them "warfighters," the term reflecting a rise in status within the fight-conscious wartime military. But the changes also complicated their ethical and professional status. Not only were they operating from dark, air-conditioned trailers, they were doing it from a world away, which troubled accepted notions of battlefield heroism.

The insecure crews welcomed the rise in professionalism that accompanied the move from observing to killing. To succeed, however, they would need to find a new place within the military social and command structure, force their machine to do things it was not designed to do, and forge ever-closer links with their comrades. From their new position within secure air bases on U.S. territory Predator crews now had access to a wealth of data and bandwidth. "Flying" the semiautomated aircraft proved less essential than network skills, cognition of data, and interpretation of video feeds.

To find out how people actually work within remote split, Tim Cullen, a PhD student in my MIT research group, conducted an anthropological study of Predator and Reaper operators, their social networks, the organizational cultures, and their evolving work experience. In a total of 180 interviews comprising nearly 160 hours, he spoke to 50 pilots, 26 sensor operators, 16 engineers, and nearly 50 others including imagery analysts, program managers, and policy makers. He also observed training missions, flew the Predator flight simulator, and observed the factory assembly of the vehicles. Cullen is himself an air force officer and F-16 pilot (with combat experience in Bosnia), so he was both able to immerse himself in Predator culture and to compare it to that of manned aircraft.

The goal of his study was to see Predator operators' work as they saw it, and to describe these vehicles and the systems they are part of, in action. The results force us to revise the public stereotype of detached individuals manipulating abstract symbols on video screens.

Cullen found operators seeking to transform themselves from "hidden automatons pushing buttons in dark rooms" to "empowered and adaptable components of respected weapon systems." They developed social relationships with broad networks of people far beyond their ground control stations. They overcame severe limitations in the systems they were given (which had been designed for another task in another time) to learn new ways of operating remotely in war. These relationships gave them a sense of presence on the battlefield that no high-definition sensor could provide.

Cullen's study returns us to the Predator control room. Ironically, despite its high-technology aura, Predator is a human-factors nightmare. It embodies old tensions about the identity of the vehicle itself, and of those who operate it. Two pilots fly it from a shipping container or building. Their control stations look less like the latest military hardware than a set of equipment racks cobbled together by undergraduate engineers the night before their term project is due.

To "fly," the two main Predator operators have to monitor sixteen different displays, interact with four touch screens, type on four separate keyboards, roll around two trackballs, flick two joysticks, and move eight levers. The main control stick and throttle are perched high on the console, making them fatiguing to hold for long periods. Manned aircraft cockpits have actually become simpler and more spare over the years, while the Predator control station has acquired screen upon screen, control upon control. It is a 1990s-era confusion of PCs, tapes, and drop-down menus.

When Predator pilots issue a command, they experience nearly two seconds of time delay before seeing it executed by the vehicle. The crew stations are not designed for comfort, making them fatigue-inducing for long missions. One 2011 study even concluded that the poor interfaces of the Predator contribute more to crew burnout than does combat stress.

It is easy to dismiss the Predator cockpit as the product of poor engineering, neglected ergonomics, and inadequate government contractors. But it actually represents the fruits of a remarkable innovative process wherein users and operators took a vehicle originally designed for a completely different task and transformed it into a global system for conducting remote warfare.

In the Predator's control van, pilots sit next to the sensor operator, at once more and less than a copilot. Sensor operators are enlisted personnel, with low rank compared to the officer pilots, but Cullen found they often felt they had the best job in the air force. With minimal training they got to control expensive equipment and unleash high-tech weapons.

Air force pilots have an uncomfortable history with this second position. Part of the fighter pilot mystique is the ability to fly alone, in total command. Over the years, only a few air force fighters have had a back-seater to operate radars and sensors (such as the Vietnam-era F-4 Phantom). It has always been a lower-status position, known as a "gib" (for "guy in back"). When the F-15 fighter was introduced in the 1970s the "guy in back" was eliminated and the job of operating the radar was incorporated into the pilot's tasks. Ironically, a later version of the F-15, the F-15E, introduced in the late 1980s, restored a back-seater, whose tasks resemble that of the Predator sensor operator. Yet the F-15E back-seater is an officer, whereas the Predator sensor operator is enlisted, often just months out of high school, leading to a different power relationship with the pilot.

In single-seat fighter aircraft, the infrared imaging sensor in the targeting pod is operated by a few buttons on the right-hand joystick and displays its image on a small screen. Fighter pilots assigned to Predator sometimes wondered how the air force could spend valuable resources training someone "to do the same job they did countless times with their thumbs and a pinky finger." Nonetheless, Predator's enlisted sensor operators experienced a high degree of immersion, and the shift of skills toward interpretation of imagery favored their expertise. "For every disgruntled [Predator] pilot hanging on for two years until he can get back into the air," ran a popular account in 2001, "there are ten sensor operators loving life."

One F-16 pilot turned Predator pilot describes the new skills required to operate in remote split. "I had TV screens all over the place—six of them to be exact . . . we had phones." He had to teach himself how to write down attack orders while flying the aircraft and talking on the phone. "When I was in my F-16, I could look out over the canopy rail and see what was going on," he recalled. "What I have to do in the Reaper is I'll take the coordinates and put them in the [sensor] pod and I will swing the pod to those coordinates."

In remote split the aircraft itself is stationed in the theater of war. A "landing and recovery element" of about fifty people maintains the aircraft, and a local pilot in a ground station taxis the aircraft and "flies" the takeoff using a control stick via a local radio link. Soon after takeoff, however, the aircraft is put on autopilot and begins to fly a preprogrammed set of waypoints. Crews then coordinate via text messages or phone call to transfer control through a satellite link to remote operators back in the United States (initially at Indian Springs, Nevada, renamed Creech Air Force Base in 2005, and more recently at other locations).

This remote "mission control element" comprises about fifty people, including ten pilots and ten sensor operators. A "mission intelligence coordinator" sits in a nearby trailer, observes linked displays, and helps the crew understand their place within the larger operation.

From a remote split node, Predator video can now be widely distributed across the network, sometimes even going back across the world. Despite these high-tech links, nearly a hundred other people still have to process, examine, and distribute the data. In addition, rooms full of analysts, such as at the "distributed common ground station" at Langley Air Force Base in Virginia, have sprung up just to catalog and analyze the rivers of data and video now coming in around the clock. While remote split certainly saved manpower and costs compared to deploying this host in the field, it is not difficult to see why the air force soon abandoned the term "unmanned" for the Predator system, preferring instead RPA—remotely piloted aircraft.

Remote split also altered the crews' work. Predator crews in the war zone, once the vanguard of new technologies in warfare, now became relatively routine operators doing repetitive tasks (although they still handled some missions locally, such as those monitoring the security of their own bases). Experienced crews at Creech (or in Missouri, Guam, South Korea, or a number of other bases) could run missions in Iraq, Afghanistan, or elsewhere with a simple tweak of a network address.

Moreover, the remote crews evolved as they needed fewer traditional flying skills. By 2004, those operating in remote split no longer needed to learn how to take off or land the aircraft. These two tasks, essential to pilots' skills and identities, had occupied almost a third of the training syllabus, so eliminating the requirement lowered training requirements and helped the air force fill Predator seats (it now costs about one tenth as much to train an RPA pilot as it does a pilot of manned aircraft). But eliminating this classic manual pilot skill further distanced Predator crews from the traditional profession of "pilot."

The pressures of war also lowered the crew's inhibitions about modifying their system. As the Iraq war began, they began to innovate and tinker with the ground control station. From 2003 to 2006 the ground control stations grew and evolved rapidly. Because of the corporate proprietary nature of Predator, operators still could not get inside the General Atomics software, so most of the innovation involved adding things—open-source software, new displays, new equipment bolted onto racks—rather than changes to the core code provided by the manufacturer. By 2005 Predator stations hosted six additional LCD displays beyond those delivered from the company.

Among the new equipment were laptops and PCs equipped with virtual whiteboards and windows into classified military networks. FalconView, for example, is a piece of open-source software created by fighter pilots to plan and track their missions that was widely used within the air force. Predator crews at first added a FalconView display to mirror the display of the mission intelligence coordinator in another room. Eventually, there would be two FalconView displays, each with its own keyboard and trackball, one for the pilot and one for the sensor operator, to show their missions on customized screens. Predator crews programmed their own FalconView add-ins that could overlay Predator video feeds, or navigation tracks, onto maps and satellite images.

Another key user innovation—and a source of several additional screens—was mIRC, or mIRC-chat, a piece of open-source software that allows simple text-based chatting on Windows PCs within the military network. mIRC was becoming popular with the U.S. military in general as it moved toward "network-centric" operations, even extending to ground troops in the field. It proved a useful channel to connect Predator crews in remote split with intelligence analysts and others across the network.

The chat format of scrolling text messages is well suited to certain aspects of Predator operations: it does not require constant attention like a voice telephone does, and it also keeps a record of transactions. Chat rooms became so important that, during missions, Predator crews would routinely monitor eight to twelve separate conversations simultaneously, sometimes as many as twenty, connecting them to imagery analysts, commanders, even lawyers and generals, in the air and on the ground.

In the words of David Deptula, the air force general who oversaw much of Predator's evolution, remote split "allows RPA essentially to carry around their own command and analysis center and legal counsel as an integral part of their payload." If crews are busy with other tasks, they can ignore the chat and then turn their attention back to it when workload permits. As crews will often say, "The Predator pilot is never alone."

But chat has the problems with online communication familiar to any of us. Messages in chat conversations refer to usernames and not to real names, and because of the anonymity, in one operator's words, "people become sarcastic, bullying assholes." Chat is also slow, and has some time delay. "If I see the guy walk out of the house with an AK-47 and I know he is a valid target," one operator observed, "it takes way too long to type in mIRC, "Hey dude, this is definitely an AK-47, lets gin up the JTAC [Joint Terminal Attack Controller] and strike this guy before he drives away." Also, when a lot is happening, the chat windows can scroll too quickly for crews to follow and absorb their information.

Its limitations notwithstanding, chat improved the Predator crews' sense of presence, though with some unusual effects. The relatively anonymous messages tended to erase distinctions of rank, providing a certain social leveling of the conversation. But that forced the crews to read deeper into the messages.

Comments from unknown users could appear at critical times—while preparing to shoot a missile, for example, users reported seeing messages like "concur" or "shoot now" or "ABORT ABORT" or even "kids." It could seem like everyone on the network had a vote in the final action. In the heat of battle, Predator crews had to interpret not only the images and behavior of the people in their crosshairs, but also the social signals coming over the network.

Remote spilt, while embedding Predator operators in large net- works, did not alone solve the problem of their isolation. In fact it might have removed them to such distance that they would lose all sense of presence by virtue of their not living in the theater of war, and sleeping alongside their comrades in harm's way while sharing their daily risks. Yet the new connections actually enabled social and organizational changes that would bring a new sense of presence to these far-removed crews.

Most Predator crew members had not been to Iraq or Afghanistan, and even those who had been there had little experience outside the perimeters of air bases. Yet here they were, watching. "Once everybody figured out how to use Predator/Reaper and what they could be used for, they wanted it for everything," one pilot recalled. "They wanted us to look at the friendlies. They want us to stare at their own FOB [forward operating base] for security. They want us to stare at a mountain pass twenty-four hours a day to see if anybody comes through a certain pass. They want us to stare at a market to make sure there is no suspicious activity. They want us to scan the road that they will drive tomorrow for IED [improvised explosive device] activity."

This new position, a mix of apparent (if illusory) omniscience and voyeurism, pressed the crews to call on new skills as they observed insurgents and civilians. Cigarettes, it turned out, glowed bright on the infrared camera, marking certain kinds of gatherings with notable blooms. What constituted threatening actions? What was abnormal? These were social judgments, evaluations and guesses of behavior and intent, for which the crews had no training, and which no automated system today can make.

To inform these judgments, remote split operators wanted more information about the situation on the ground than the aircraft itself could relay, the all-important context of what appeared on their screens. They had to extract this information from people in their networks, including the troops on the ground whom they were supporting, indeed fighting virtually alongside. Yet those closest to the fight had other things to worry about; they just wanted to see what was going on around them.

Several years into the Iraq war, a pilot we'll call Colonel Smith took over as commander of a Predator squadron. He had flown the A-10 Warthog, the ungainly air force jet specialized for ground attack. Smith found his Predator squadron a "depressing place" with low morale among the operators. The pace of operations had doubled in a couple of years, and crews were exhausted. More problematic, the crews often did not know much about the missions they were flying. The ground troops would simply not tell them what they were doing, or why they were looking for something in particular. Instead, the Predator crews were told to fly some distance away and point the ball at a certain place and keep their eyes on a building or a person; then they might be ignored for a couple of hours. Predator crews began to feel like "chat-activated sensors" and lost their motivation when disconnected from "the big why." Smith found they continually needed to "fight for context."

Smith traveled to Iraq to speak with the troops. He tried to convince them that Predator crews had expertise on the best way to employ their system. Because of their presence within U.S. networks, he argued they had access to a great deal of information that the ground troops lacked, and could share some of it with those on the ground through the radios mounted on their vehicles. Smith argued that the more his crews knew about the mission under way, the higher quality of support they could provide. Smith appealed to the ground commanders to trust the Predator crews and give them the information they needed.

Smith built personal relationships with the commanders his team would be supporting. After his trip, those same troops returning from Iraq would visit the remote split Predator squadrons to meet them and observe operations. They began providing more information and details about what they were looking for. Morale improved as expertise shifted back through the network to the Predator crews.

This shift toward more active, professional Predator crews quickened with the introduction of Predator B, or Reaper, which began to appear around 2004 and grew to nearly half the Predator force by 2015. Reaper was a larger, highly improved Predator that could fly twice as high, twice as far, and twice as fast, with greater numbers and variety of weapons, even though it was still flown with the same limited ground station.

Again, the technological change brought professional shifts. Reaper squadrons were made up mostly of fighter pilots, who thought they worked in a more "tactical" way than Predator operators, who were more likely to be transport pilots. They sought a "tactical mindset" which meant a closer working relationship with those in the field. Reaper pilots sought to change the language of their work: they did not "operate a sensor ball" but rather "flew a targeting pod"; they did not "conduct

ISR" missions but flew "nontraditional" intelligence missions, just like fighter and bomber crews did.

Predator crews developed close relationships with JTACs (joint terminal attack controllers)—the air force teams embedded in army units that could direct weapons fire from Predator or other aircraft. Friendly troops no longer saw the vehicle as a disembodied eye in the sky feeding a faceless intelligence bureaucracy, but as part of their system of field support.

Along with the imagery, voice contact with JTACs in the combat environment also helps Predator crews feel present. During combat, they hear gunfire and stress in voices as troops run for cover. JTACs might talk in hushed voices while hiding, and Reaper crews will find themselves whispering to each other too. Predator crews often reported that their interactions with those on the ground, especially the JTACs, were the most satisfying parts of their jobs.

Nonetheless, this sense of presence has limitations, which draw our attention to its artificial nature. To begin with, and perhaps most obviously, the presence is through an American lens. The video, chat rooms, and voice links that connect Predator operations to the battlefield see the world through one side. Evoking the intentions, identities, even sometimes the ages and genders of people from the local populations, is much harder to do, and has accounted for civilian casualties from Predator operations.

In a particularly tragic episode, Predator crews jumped to the conclusion that an early-morning truck convoy carried teams of insurgents. In fact, it was a family on its way to a wedding; the mistaken American attack killed twenty-three noncombatants and injured sixteen. The error stemmed not from the limited video-screen resolution that made it tough to distinguish guns from shovels, but from confusion and a collective desire among the operators to see what they wanted to see, and what they presumed was happening.

"Technology can occasionally give you a false sense of security," concluded Air Force Major General James O. Poss, who oversaw the air force investigation into the incident, "that you can see everything, that you can hear everything, that you know everything." Afterward, the air force prohibited Predator crews from using the term "military-age male" because of its implication that all adult men were enemies.

Even friendly intentions can be difficult to interpret. In one episode, a group of insurgents ran out of a burning building, and one of them lay down on the sidewalk, hunched over. American ground soldiers responded to the incident. Predator operators watched an American soldier walk up to the insurgent, step back, raise his rifle, and shoot him dead. The Predator team filed a grievance for the soldier's violation of the laws of war, initiating an investigation. The inquiry revealed that the insurgent was wearing a suicide vest and was about to detonate it when the solider shot him. "By not understanding that you have a two-dimensional, twenty-degree field of view of what is going on in life," one pilot came to realize, "you think you have situational awareness when in fact, you didn't hear the commands."

Such mistakes, of course, including callous disregard for human life, also happen with warriors physically present on the battlefield. Accurate numbers are difficult to find, but those who have looked at civilian casualties in remote split have found them to be about comparable to those from other weapons. But with Predator the mistakes are recorded on video, and draw on cultural fears of robotic attack.

Much has been said and written about Predator pilots' physical detachment from the battlefield. They go to work each day, conduct a war, and return home to ordinary tasks of daily and family life. Actually, this phenomenon is not new. In fact, aircrews have long lived at a distance from the battlefield and in more comfortable surroundings than their brethren fighting on the ground. One difference here, of course, is that the Predator crews are not risking their bodies by flinging them high above the fight, with at least a finite chance of ending up down below on the enemy side of the fray.

But Predator operators also differ from traditional aircrews in that they experience continuous shift work, for years on end. The tempo and pace have their own psychological and social effects. "It's a three hundred sixty-five-day job," one Predator pilot remarked. "You can be in the squadron bar on Friday night—there's dudes that are working." The nature and pace of the operations means that "you can't get the entire squadron together. . . . You definitely can't hang out with guys."

A study of stress and burnout among Predator operators found their problems were due to the relentless, high pace of operations, the troubling difficulties of the user interface, and the uncertainties in career advancement more than to the moral stresses of witnessing and performing remote killing.

Nonetheless, a small number of Predator operators and imagery analysts have been speaking out in the press, relating their own experiences of war based on the camera's intimate vision. "I knew the names of some of the young soldiers I saw bleed to death on the side of the road," imagery analyst Heather Linebaugh wrote in *The Guardian* in 2013. "I may not have been on the ground in Afghanistan, but I watched parts of the conflict in great detail on a screen for days on end. I know the feeling you experience when you watch someone die." She reported being haunted by a lingering sense of doubt about the accuracy of the identifications she made that resulted in other peoples' deaths.

Brandon Bryant, a Predator sensor operator from 2007 to 2012, reported to *GQ* magazine that while he was mocked by buddies for being an armchair warrior, he struggled "to square the jokes with the scenes that unfold on his monitors." He reported feelings of helplessness, disconnection, and powerlessness and of suffering post-traumatic stress disorder (PTSD) after his Predator duties. To friends who mocked his experience on his Facebook page, he asked, "How many of you have killed a group of people, watched as their bodies are picked up, watched the funeral, then killed them too?"

A 2013 air force study found that RPA crews had the same levels of mental health diagnoses, including PTSD, depression, and anxiety disorders, as aircrews in the field. The equivalence itself is surprising, as it compares crews at risk in a war zone with those living at home with their families. An April 2014 study by the General Accountability Office (GAO) reported staffing problems and low morale, poor working conditions, uncertain career prospects, and personnel shortages among RPA pilots due to the "stigma" of flying unmanned aircraft. Pilots reported that a major stressor was their lack of clarity about when they would return to careers in manned aircraft, and the vague and all-encompassing nature of being "deployed on station" at bases within the United States. Many said they would prefer to deploy to a theater of war for six months, with a clear end to the assignment, rather than being deployed on station for three years or more of six-day-per-week shifts. RPA pilots, reported the GAO, are leaving the air force at three times the rate of pilots of manned aircraft.

The air force responded by pointing out that RPA pilots are really not the best pilots. "Let's be honest, when people dream about flying," an air force spokesperson told ABC News, "people in this generation didn't grow up and say, 'I want to fly an RPA.'" This insult from the air force hierarchy, in the face of the acknowledged importance of

Predator and its crews, reflects a service still unsure of the status of this new, technologically mediated experience of war.

> *They shot down a Predator, that's one less slot for me.*
> *They shot down a Predator, and it filled my heart with glee.*
> *You know the sucky thing is, the Air Force is just building more of them . . .*
> *They shot down a Predator and I wonder how that feels*
> *For that operator, who lost his set of wheels.*
> *It must feel so defenseless, like clubbing baby seals.*
> *They shot down a Predator, and I wonder how that feels.*

DOS GRINGOS, "THE PREDATOR EULOGY,"
LIVE AT THE SAND TRAP

"The Predator Eulogy," a song written by two fighter pilots known for their irreverent music, wryly exposes how the status of the Predator is far from settled within the air force while acknowledging the anxiety it generates ("one less slot for me"). The song probably refers to a number of incidents where Predators lost their radio links and had to be shot down by American planes to prevent them becoming a hazard to other aircraft. One even hears reports of these pilots painting Predator icons on the sides of their aircraft as symbols of aerial victory, until being forced to remove them by air force higher-ups.

"After I finished laughing," one F-16 pilot reported when he was asked to fly escort for a Predator, "I refused and went on with more important business." In typical language denigrating RPA operators, he described RPAs as "becoming fashionable with the bespectacled computer-screen officers living in fortified operations centers." Many pilots point out the benign air environment in Iraq and Afghanistan, free of threatening air forces, arguing that RPAs would not be effective in a situation with missile defenses, enemy aircraft, and antiaircraft artillery. "In other words, a war."

In 2011, Dave Blair, an air force officer studying for a PhD in International Relations, pointed out the basic contradiction in U.S. Air Force culture in an article in an air force journal. "When a manned aircraft with two spare engines scrapes the top of a combat zone, well outside the range of any realistic threat," Blair asked, "why do we consider that scenario 'combat' yet deem a Predator firing a Hellfire in anger 'combat support'"?

Blair recalled a barroom argument with a pilot of the F-22, the air force's advanced fighter whose very cockpit resembles a Predator control station. The pilot told him "fighting a war via video teleconference isn't very honorable." Blair noted that "we might say the same for firing a missile beyond visual range from a fighter cloaked with stealth technology [the F-22 pilot's job]."

In print, Blair described the air force's sending "conflicting institutional messages" as a source of Predator crews' morale and personnel problems. He called for the service to recognize that Predator crews were having significant effects in combat, albeit with reduced risk to their bodies, and to award them combat medals as a means to improve their sense of accountability.

The response was immediate, vociferous, and personal, declaring Blair's paper "ludicrous at best and insulting at worst." Had he had the experience of combat, one commentator railed, "maybe he would understand better while under direct fire." Actually, Blair had served as a pilot of both a C-130 gunship and a Predator, in both Iraq and Afghanistan, in his words, "both physically and through telewar." Still the critics pointed to the essential value of direct experience: "To bill their [RPA] controllers as being in combat just isn't true. One must experience it firsthand." "No way is a UAV pilot [sitting] in a box," another opined, "facing the same risks as pilots operating manned aircraft over enemy territory."

Such heated exchanges, undoubtedly repeated in countless hangars, operations centers, and bars, have forced the air force to examine its notions of "airmanship." What is the necessary relationship between flying an aircraft and exercising command? How does the element of risk define military professionalism, or even the legitimacy of killing in war? What are the risks to pilots flying manned aircraft high above the battlefield in uncontested airspace, other than the unlikely occurrence of crashes due to mechanical failure? Perhaps most starkly: How do officers balance their commitments to the profession of aviation and to the larger military mission?

In 2009 the air force created a new career classification for RPA pilots, dubbed 18X, with a new form of undergraduate training, that, over time, is supposed to supply the lion's share of crews. It remains to be seen whether this new designation will earn legitimacy in a military hierarchy organized around manned aircraft.

In February 2013, as though responding to Blair's suggestion, outgoing defense secretary Leon Panetta announced a new decoration, the Distinguished Warfare Medal, that does not require an act of physical valor and combat risk. It is designed specifically for those in drone or cyber-warfare who would not be eligible to receive other medals. It sought to recognize "extraordinary achievement, not involving acts of valor, directly impacting combat operations or other military operations." The new medal would rank above the Bronze Star, given for performance in a combat zone, and below the Distinguished Flying Cross.

This ranking raised the ire of veterans' groups, which derided the

Distinguished Warfare Medal as the "Nintendo Medal." How could any award recognizing physical service in combat, veterans asked, be ranked lower than that of a remote operator? "The contribution they make does contribute to the success of combat operations," Panetta responded, "even if those actions are physically removed from the fight." The Military Order of the Purple Heart joined in condemnation, stating that "to rank what is basically an award for meritorious service higher than any award for heroism is degrading and insulting to every American Combat Soldier, Airman, Sailor or Marine who risks his or her life and endures the daily rigors of combat in a hostile environment." In response to these objections, Panetta's successor, Chuck Hagel, ordered a review of the medal as soon as he took office. Within weeks, the new medal was canceled.

In 1862, a crew member of the new ironclad warship USS *Monitor*, staring at his new, ominous iron home, wondered "whether there isn't danger enough to give us glory." The new technology, with its apparent ability to protect him from the risks of enemy shellfire, changed his definitions of heroism. Herman Melville, upon visiting the *Monitor*, wondered whether "warriors are now but operatives," or factory workers. Predator pilots inherit these anxieties, updated in a world of global networks, cognitive systems, and knowledge work.

The Predator cockpit is more than a place where pilots fly airplanes. It is a node that brings together the people and information required to fight a modern war, from aircraft designers, programmers, pilots, sensor operators, troops on the ground, and analysts who consume the data. It does so within a context of distant wars with shifting goals operating under fragile political consensus. Presidential power, congressional oversight, intelligence agencies and the public are all part of Predator's network. Warfare is always a social experience, and even the enemy and the civilians with whom they mingle have a presence here. The job of the Predator crew is to sort through these conflicting representations, conflicting social agendas, and conflicting professional identities to achieve their goals, which are often unclear. Through the flashing screens they enter into a different world.

Yet, in the words of a Predator pilot at a recent air force symposium on the "future operator," "UAS pilots/operators have an identity crisis." On the one hand, they are mocked and reviled as armchair warriors, unable to risk their bodies in heroic conflict. On the other hand, they are stressed because their work is in such high demand, work that gives them a novel, telescopic view into the horrors of a new and unfamiliar kind of war. Like the crew of the *Monitor*, they wonder whether there isn't danger enough to give them glory.

In the civilian world, danger and glory do not press with the same acuity as in warfare. But professional dignity, social status, and changing tasks are similarly at stake in modern workplaces. When confronting new machinery, we are all Predator operators.

CHAPTER 5

Space

AS AN EXTREME ENVIRONMENT, SPACE IS SUPREMELY HOSTILE to human life—vacuum, radiation, cold, and sheer distance push the boundaries of human experience. In space, without protective suits or vehicles people will simply freeze, burn, or explode. It is a world of long distances and hostile conditions, and fraught with high political stakes. From its beginning in the twentieth century, spaceflight has raised the question of what kind of presence, and what kinds of machines, best convey human experience.

Nearly fifty years ago, when Neil Armstrong landed the lunar module on the moon, he had both a HUD and an autoland. The HUD was an early, passive design—a series of angle markings etched into Armstrong's window. The onboard computer read out a number on an LED-like display, and if Armstrong positioned his head correctly he could look through that angle marked on the window. Directly behind the indicated number, then, he saw the actual spot on the moon that the computer was flying him toward.

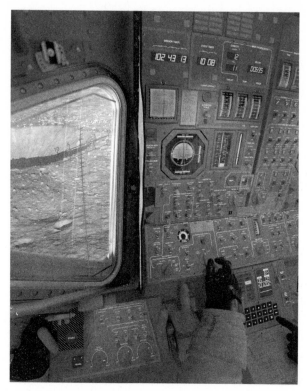

Computer graphic depiction of Neil Armstrong's view out the window from a few hundred feet above the moon during the Apollo 11 landing. Armstrong's right hand is reaching up to turn off the automatic targeting and autoland features to land in a semimanual mode. Note the Apollo Guidance Computer on the lower right, which could indicate an angle for Armstrong to look through the passive HUD inscribed as a reticle on the window.

(IMAGE BY JOHN KNOLL ACCORDING TO THE AUTHOR'S CONCEPTION, WITH RESEARCH HELP FROM PAUL FJELD)

If he didn't like the designated area because of rocks or craters, he could give a momentary jog to the joystick in his hand and "redesignate" the landing area fore, aft, or right or left, almost like moving a cursor with a mouse to tell the computer where to land. The computer would then recalculate the trajectory it would use to guide the landing, and give him a new number for where to look. Armstrong could redesignate the landing zone as many times as he liked in a human/machine collaboration that allowed the two to converge and agree on the ideal landing zone, whence the autoland would bring the craft safely down.

But Armstrong never used the autoland. On Apollo 11, he didn't like the spot the computer had selected for him because there was a crater, and rocks were in the way. Rather than redesignate, a few hundred feet above the moon's surface he turned off the automated feature and landed in a less automated mode. (His actions were echoed by Luke Skywalker in the original *Star Wars* movie: he also turned off his computer and relied on The Force to guide his aim to destroy the Death Star.)

My earlier book explored this and the five Apollo landings that followed, when every single commander turned the autoland off at about the same height. Most of them said they turned off the digital aids because the landing spot had rocks or craters, though the computer was perfectly capable of allowing them to redirect the landing.

The space shuttle, too, had an autoland system. After blazing through a hot reentry from space, the shuttle could follow a specially designed microwave beam and its computer would fly the craft all the way to the ground, with the pilots only monitoring and never touching the controls (though they still had to manually deploy the landing gear and the drag chute). The space shuttles flew 135 missions between 1981 and their retirement in 2011. On every single flight, the shuttle commanders turned off the automation well before landing and hand flew the vehicle to the ground.

The closest the autoland came to landing the shuttle was on the third shuttle flight when, as a test, astronaut Jack Lousma allowed the autoland to fly the vehicle as close as 120 feet to the ground. In high winds while landing at White Sands, New Mexico, the software did not control the approach perfectly, and the transition from automatic to manual resulted in some unplanned movements, including a large "wheelie" once the rear wheels touched down. The software was improved, but the autoland system was never used again in the shuttle program.

It's easy to ascribe these actions to pilots' rejection of automation in favor of their own hands-on skill, a trade union resistance to technological change. Undoubtedly professional pride played some role, but focusing only on that factor narrows our vision.

I asked former shuttle commanders why they didn't use the autoland even though they had trained for it in simulators and in simulated landings in the shuttle practice aircraft. They responded that if something had gone wrong and they needed to intervene manually, it would be too difficult and disruptive to "get into the loop" so near to the landing (also, the sensitive ground equipment required to guide the autoland was not available at all of the shuttle's backup landing sites). Given that you've got only one chance to land a shuttle, you might as well hand fly it in. Think of the Air France 447 pilots' surprise when the machine checked out.

Yet, like the EuroAir pilots mentioned earlier, the shuttle pilots did follow a guidance cue and a flight path vector. In fact, those data came from the autoland system, which was not really turned off but only prevented from issuing steering commands to the shuttle. Through the guidance cue, the autoland algorithm was, in effect, commanding the pilots to steer the shuttle.

Apollo astronauts and shuttle commanders, carrying the weight of their country's fortunes and risks to their own skins, made the same choice that EuroAir was making about their relationship to the machine: rich interaction, with visual feedback, rather than automated "on top of the loop" supervision.

The shuttle's neglected autoland is but one grain of the ever-shifting sands of human presence in space flight. In 2011, after forty years of human presence, the retirement of the space shuttles put Americans at a turning point that challenged their imaginations. Both actual and virtual human flight in space had worked spectacularly well, though marred by disaster.

Looking to the future, observers asked whether people still need to venture into space or whether all that they do can be accomplished remotely from the ground. At the same time, when contemplating the shuttle's legacy, what stands out is not only the scientific accomplishments, but also the human manipulations. The repairs to the Hubble Space Telescope and the construction of the International Space Station seemed to reinvigorate human presence of a traditional kind: fixing and building things.

Simultaneously, a series of small, robotic rovers on Mars provided an intense, first-hand sense of presence on the planet, both for their scientific teams and for a large public audience, despite limited data bandwidth and long time delays. How did these scientists feel "present" in remote places? How do people work on Mars?

Let us look at these well-known instances of spaceflight, both human and remote, to begin to answer these questions. The service and repair missions to the Hubble Space Telescope and the remote exploration of Mars through robotic rovers highlight human presence in space, but in rather different roles: in one, the human as repairman or construction worker, exercising manual dexterity, and in the other, the human as explorer, exercising scientific judgment.

In both, experience and skill travel across vast, interplanetary networks as humans learn to work outside our earthly envelope. The rise of robotics had a profound influence on the missions to repair the

Hubble Space Telescope, and hence on the construction of the space station. Remote rovers on Mars enable scientists and engineers to be present on another planet and do their scientific work there on a daily basis.

The Hubble Space Telescope, though it has been in orbit around the earth for more than twenty years, was repaired and upgraded by human hands working in the large bay of the space shuttle five times over a sixteen-year period. At first glance, the missions seem the ultimate expression of manual skills, exercised by fine craftsmen in orbit. But the tasks were also intimately connected to robotics. Before examining the further reaches of remote presence on Mars and beyond, then, it is worth considering how "human" spaceflight is intimately bound up with machinery, and how working with robots is a primary skill for human astronauts.

The story of a space telescope is itself a story of the changing experience of work for astronomers. As the ultimate remote explorers, they never visit the sites they study. Yet they do have a tradition of focused observing sessions in isolated settings, high on remote mountains like Mauna Kea in Hawaii, or even just late at night on the roofs of buildings in many locations. Today astronomers often observe remotely, issuing requests from their offices for telescope time, then receiving images and data over the Internet. The space-based telescope is but the most extreme example of astronomers being displaced from their sites of observation.

The idea of a telescope in space has a long history, going back to German rocket pioneer Hermann Oberth, who proposed the idea in 1923. In 1952, his protégé Wernher von Braun teamed with artist Chesley Bonestell to publish an influential series of articles and images in Collier's magazine, laying out a vision for humans in space. Von Braun included an orbiting space telescope, operated robotically except for

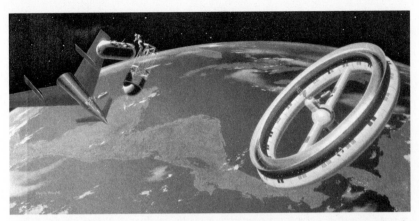

Chesley Bonestell image of future space operations, 1952, depicting a space shuttle (left), a space station (right), and a space telescope (center) with an astronaut servicing it to change the film.

periodic visits by astronauts to change the film. Von Braun failed to anticipate advances in electronic imaging, but the idea of humans as servicers of remote telescopes indeed came to pass.

Still, early conceptions of the Hubble involved in-flight servicing in a robotic mode. The shuttle would rendezvous with the telescope, grapple it with its manipulator arm, and then place it on a rigid mount in the payload bay (much as it did on the actual servicing missions).

The vision, however, did not require astronauts to don spacesuits and manipulate the telescope's instruments directly. Rather, the arm would automatically replace the modules, sliding each one out, stowing it, and putting another one in. All components would be well known in advance, and mounted in a rigid structure, so the task seemed feasible even for 1970s robotics, akin to an automated task on an assembly line.

The human involvement would come from within the shuttle, where an astronaut would command the arm to perform preprogrammed tasks. "All the astronaut has to do is sit in a bay and push a button," Frank Cepollina, NASA's father of satellite servicing recalled, "and the arm would change out the modular instruments." A large mockup was created for this purpose, with instruments pulled out and replaced robotically. This version of the task helped spur development of the shuttle's Canadian-built remote manipulator arm, which came into being relatively late in the shuttle's development.

Much of this original vision indeed survived into the Hubble repair scenarios—the rigid mounting of the telescope, the manipulator arm operated by a human inside the shuttle. But the crucial additional elements are the astronauts in space suits working in the payload bay, frequently attached to the end of the arm, sometimes with nearly their entire bodies snaked up inside the instrument.

Hubble was NASA's first satellite designed to be serviced by human hands. It reached orbit in 1990 and the occasion soon arose. Due to a series of miscalculations and missteps by engineers on the ground, minute imperfections had been mistakenly built into both the telescope's main mirror and the test equipment used to evaluate it.

When it failed to deliver the expected results, the expensive telescope, brought in over budget and years late, looked like a lemon—"Right next to *Hindenburg* was Hubble," recalled astronaut Jeff Hoffman. "It was a disaster for NASA." The cover of *Newsweek's* July 9, 1990, issue ran the headline "Star Crossed: NASA's $1.5 Billion Blunder," just one of a panoply of jokes and recriminations.

The extraordinary series of Hubble repair missions began with the need to replace the telescope's faulty optics. NASA assembled a repair crew that brought a depth of spaceflight experience to the job. The two lead spacewalkers were Jeff Hoffman and Story Musgrave. Hoffman, an astrophysicist with a PhD from Harvard, had flown three previous missions. Musgrave is a man of varied background: a Marine aviation

mechanic, pilot, and physician. His experience as a surgeon lent authority to the precision manual nature of the task. As with a patient, they
would be opening up a body, replacing parts, fixing problems, doing
some essential upgrades, then closing it up and sending it back on
its way.

Before the mission, NASA administrator Daniel Goldin called the
crew into his office and told them the future of the agency rested on their
work. The proposed space station was just making its way through Congress, with uncertain support; NASA must prove that it could recover
from the Hubble mistakes and perform complex, high-risk manual tasks
in orbit. The station would take a great deal of manual labor to assemble,
but if NASA could not redeem itself with the Hubble repair, Congress
would not trust it with the much larger project. The apparently mechanical job of fixing an instrument to peer into the origins of the universe
had deep political implications on earth.

To simulate weightlessness, the crew spent endless hours practic-
ing in NASA's large underwater practice pool. Procedures were so complex that Musgrave and Hoffman couldn't fit all the steps for them on
their wrist-mounted checklists; their colleagues inside the shuttle had
to read them instructions over the radio. Hoffman and Musgrave refined
and rewrote the manufacturer's procedures, choreographing their every
move and accounting for untold contingencies.

The shuttle *Endeavor* lifted off in December 1993, embarking on the
most ambitious mission of the shuttle program to date, with five space-
walks scheduled and another two possible if necessary. Once in orbit, the
day after liftoff *Endeavor* closed in on the Hubble while the crew checked
out their equipment. When they spotted the stricken telescope, another
problem immediately became apparent: its solar panels, which had been
acting up, were physically damaged. The next day, in preparation for the

repair, ground controllers in Maryland commanded the Hubble to shut down into safe mode and stow its antennas.

Under computer control, the shuttle thrusted closer and closer to the telescope, until commander Richard Covey took control and manually flew the vehicle about thirty feet away from the instrument, which seemed perfectly stationary. Hoffman remembered it as "really kind of a magical thing . . . you look at this big telescope out there, and it's just floating motionless above the payload bay, and yet we're both moving at 18,000 miles an hour."

Astronauts Jeff Hoffman (stripes on legs) and Story Musgrave during the STS-61 space shuttle repair, 1993. Note Hoffman's legs strapped into the shuttle's remote manipulator arm.

(NASA JOHNSON SPACE CENTER)

Swiss Astronaut Claude Nicollier operated *Endeavor*'s arm to grab the telescope and bring it to a secure berth in the shuttle's cargo bay. "We have a handshake with Mr. Hubble's telescope," Covey called over the radio. Then the captured telescope was latched into the bay. The crew inspected their quarry with a remote camera on the end of the shuttle's arm.

On day three, Musgrave and Hoffman began their first spacewalk by suiting up. Their tools hung from carriers on their chests and on a long device they called a "fish stringer" because of its multiple hooks.

Once outside in the vacuum of space, festooned with tools and mechanical appendages, they began setting up their work area inside the shuttle's cargo bay. They organized the bay like any craftsman's workbench, with an array of tools neatly set alongside the work pieces, some required and some there just in case they were needed. "We had hammers, hacksaws, crowbars," Musgrave recalls. "Luckily we didn't have to use any of that rough stuff, but we were ready."

Hoffman then put a foot restraint onto the end of the shuttle's arm, and strapped himself in by his feet. Nicollier, operating still from within the shuttle, then "flew" the arm, and his colleague Hoffman, around the telescope for further visual inspection. Early in the mission, Hoffman would tell Nicollier how to move him; as the mission progressed Nicollier learned to anticipate Hoffman's actions and to move him before he asked, silently coordinating.

At one point, in what became known as "the great screw chase," a small screw escaped a tool bag and began to float outside the cargo bay—a potential danger to the Hubble or the shuttle's delicate mechanisms. Nicollier moved Hoffman far outside the cargo bay to grab it, but the arm moved too slowly to keep up. Shuttle pilot and backup arm operator Ken Bowersox went to the computer, fiddled some parameters, and speeded up the arm. Hoffman caught the screw.

Soon Musgrave and Hoffman opened the large doors on the side of the telescope, and Hoffman climbed in. Working like a mechanic inside a car, Hoffman removed several units containing failed gyroscopes and replaced them with new ones.

At the end of the day, while Musgrave moved on to another task, Hoffman tried to close the doors, something he had practiced in the pool. "So I closed the door, I turned the latch, and I went up to the top to throw the bolt." But the door did not close. He called Musgrave, who came over and found he could hold the top of the door but couldn't throw the bolt. "It was basically a five-hand job," Hoffman remembers in frustration, "but we only had four hands."

To solve the problem, they called on their extended network, describing the problem to engineers on the ground and taking and downloading photographs. The distributed team developed a solution—to use a payload constraint device, a kind of cargo strap, to ratchet down the door. It finally closed. Their spacewalk lasted nearly eight hours, the second longest in NASA history. Hoffman found it mentally but not physically exhausting.

The next day, astronauts Thomas Akers and Kathy Thornton replaced the telescope's damaged solar cell arrays. They also gingerly installed COSTAR, the corrective optics assembly that would redress Hubble's core optical flaw.

On the sixth day, working at night in the glow of a spotlight from the shuttle's payload bay, Hoffman and Musgrave replaced the Hubble's Wide Field Planetary Camera (WFPC). Again riding on the end of the arm, guided by Nicollier and helped by Musgrave floating alongside,

Hoffman removed the grand-piano-sized instrument out of the tele-scope. He placed it in a special carrier in the cargo bay and reversed the process with the new one, keenly aware that any jarring or smudge would mar all the telescope's future imagery.

Three more days of similar acrobatics replaced solar array drive elec-tronics, computers, and other components. After some testing, Nicollier again grabbed the telescope with the arm and released it into space.

Musgrave, when he talks about the mission, uses terms like "cho-reography" and "ballet" to describe the slow, graceful, and critical posi-tioning of all these elements in space and time. He considered the astronauts as fine links coupling the large human/telerobotic system: "we are just eyes and hands and we're mission control's extension." Shuttle pilot Ken Bowersox called the mission "arts and crafts for pilots and flight engineers." It was also very tense, as anything could go wrong at any moment, with expensive, embarrassing, or dangerous results.

STS-61 was a great success, enabling Hubble to become the most productive science mission in NASA's history, capturing direct views into the early universe, measurements of the size and age of the uni-verse, observations of solar systems in formation, and other phenomena that have rewritten astronomy textbooks.

The first Hubble repair was not only surgery, not just fine manual manipulation, but an interplay of human and robotic elements. There, at the end of the arm, humans stood as the "end effectors," the eyes and ears of a planetary human/robotic system.

It was a remarkable assemblage: the shuttle, with its crew of five inside, communicating with Houston via voice, data, and video; Nicol-lier operating the arm, each joint under his precise control, looking out the window at Hoffman and Musgrave, Akers and Thornton, operating on the giant remote eye under the control of another group of humans in another part of the United States. To Frank Cepollina, the mission proved what he had argued for all along: "We should never run a servicing mission without humans and robots . . . robotics and humans go together."

In three more missions following the original repair, in 1997, 1999, and 2002, astronauts continued to upgrade Hubble. They added cameras and spectrographs, swapped out decaying sensors and batteries, and generally kept the telescope abreast of the latest electronics. A final servicing mission scheduled for 2004 was planned followed by a trip in 2010 to retire and return Hubble to earth for display in a museum.

Then, in January 2003, the space shuttle *Columbia* broke up upon reentry, killing seven astronauts. All space shuttles were grounded while NASA investigated the accident and reevaluated its mission.

A year after the *Columbia* accident, NASA administrator Sean O'Keefe announced the shuttle would not do further Hubble repairs. As justification, O'Keefe cited the risks: because Hubble sat in an unusually high inclination, astronauts could not access the space station as a "safe haven" if they had a *Columbia*-like problem that would preclude safe reentry. But O'Keefe's decision came just two days after President George W. Bush announced his new "Vision for Space Exploration" to focus on a return to the moon. Low earth orbit was out of favor; space observers saw the Hubble as the first "victim" of the new Bush policy.

Advocates went to work persuading NASA and Congress to reverse the decision. In the meantime experts at NASA's Goddard Space Flight Center began designing a mission to do the final Hubble upgrades as an unmanned robotic mission. From March 2004 a team of more than a thousand people, based at Goddard but including numerous other NASA centers and contractors, spent a year working the problem, planning for a 2008 mission and carrying the project through its preliminary design reviews.

Michael Griffin took over as NASA's new administrator in 2005, skeptical of the feasibility of the robotic mission. His wariness was bolstered by a study from the National Research Council that found the robotic approach extraordinarily risky and preferred the human-servicing option. Griffin immediately canceled the robotic effort and soon announced a shuttle-based mission, known as SM4 (for "servicing mission number four") which was ultimately scheduled for 2009 on shuttle mission STS-125.

Goddard's robotic mission was never flown. Still, it's worth our attention as a natural comparison of human and robotic servicing in orbit. What's more, the planning for the robotic tasks influenced the eventual human tasks, giving us another case wherein human and remote systems evolve together.

The Goddard team found that designing a robotic servicing mission was enormously complex—especially for an instrument like Hubble, which was designed to be serviced by people and not machines. Hubble's finicky gyros, for example, were buried deep in the telescope's structure. Six astronauts, by craning their bodies inside, had changed out eight of the units on three prior servicing missions, and each time they had trouble.

Arthur Whipple, systems engineer on both the robotic mission design and the human-servicing mission, concluded that, despite their best efforts in robotics, "an interface that by design is difficult to install, will never get easy no matter how hard you work at it." It was so difficult to imagine a robot reaching inside the Hubble bay the way Hoffman had that the Goddard team finally elected to add new gyros on the outside of one of the camera packages. They also concluded that equipment that was well designed for robotic servicing was also just good design practice in general—mounted on the outside of the spacecraft, easily accessed, and simply removed. What's good for robots is also good for humans; what's deficient for humans is even more deficient for robots.

The Goddard team designed an array of exotic tools for opening doors, latching them open, and positioning other tools. These tools were later adapted for human use. The final servicing mission, SM4, flew its seven crew members with sixty-six tools from previous missions. It used more than a hundred brand-new tools originally designed for robots.

One of these exotic tools just held on to the hardware. NASA wanted to replace circuit boards deep in the Hubble's guts which, unlike earlier replaceable units, were not designed to be repaired. This required removing tens of tiny screws. For the robotic mission, Goddard engineers designed a special capture plate bolted to the outside of Hubble. On SM4, it allowed the human crew to extract the screws without their drifting away—avoiding another "great screw chase." Thus the human mission could conduct major surgery, removing more than one hundred such screws, then pull off the plate as a unit and keep the small parts contained. "Design and testing of robotic tools," Whipple concluded, "reinforced the value of specialized tools for human servicing."

The SM4 mission on STS-125 became the most productive of the Hubble servicing missions. Its success derived at least in part from tools designed for robots. It also encountered a number of unexpected situations that would have been difficult for a robot to handle. At one point astronaut Mike Massimino was removing a handrail and encountered a stripped bolt he could not remove with his hand tool. After some deliberation, he just tore off the handrail by force.

When comparing the human and robotic servicing missions, the

Goddard team quantified the differences in *time*: the robotic mission was scheduled to last seventy-three days with sixty-one days servicing, whereas the human equivalent lasted only thirteen days from launch to landing, with six days of servicing.

Human servicing, Whipple concluded, necessitates a shorter mission but is efficient, whereas robotic missions take longer but have fewer time constraints. Latencies in communications stretch out the duration by making the act-and-respond cycle of telerobotic manipulation especially slow. Whipple encountered a subtle trade-off between cost, complexity, time, and the presence of a human or robot in the loop, "a continuous and evolving structure from [human] EVA to autonomous robotics."

As impressive as these Hubble missions were, they were basically mechanical tasks, not so different from von Braun's original idea of changing out the film. The unique human capabilities came from dexterous mechanical manipulation coupled with the human body's rich awareness of its physical environment. Yet the question of time raised by the Goddard team links the space servicing missions to the much broader human activity of exploration.

Steven Squyres, the chief scientist of the Mars Exploration Rover (MER) missions, often comments on the frustratingly slow pace of their work. "It took four years to do a week's worth of fieldwork!" he writes. "It has unfolded in excruciatingly slow motion." Here he echoes Whipple's finding about the Hubble repair missions: robots are slow, especially with the inherent time delays in remote space work (seconds in the case of Hubble; twenty minutes in the case of Mars). Scientists, specifically field geologists, argue for the special human insights brought to bear by human beings in rich, real-time interaction with their environments.

This issue of time, then, returns us, via the moon and Mars, to field geology, the science that Bob Ballard learned how to do with *Alvin* and remote submarines, as a focal point for the human and robotic roles in space exploration.

Kip Hodges, founding director of the School of Earth and Space Exploration at Arizona State University, is an accomplished field geologist. He describes his work as "best done by one or two geologists in the field working alone," traveling through the terrain, working multiple hypotheses that evolve in the course of exploration as scientists stop to examine what they've encountered and replan accordingly. The ultimate goal in his work is to make a geological map that constrains the plausible geological histories of the study area. He stresses geology's essentially craft character as a science that "relies on creative thought rather than rule-based execution." To Hodges, doing field geology "seems impossible with autonomous robots."

The irony is that this science, while taking the lead in what were, typically, far-flung locations, was embattled in the twentieth century. Historian Naomi Oreskes points out that American geologists rejected the theory of plate tectonics for decades, in part because it threatened their beloved practices of field geology. Large-scale theories like plate tectonics depended on the rise of more laboratory-based, quantitative sciences. The choice of theories for field geologists was a choice between two different ways of life. Hence Bob Ballard's successful intervention in learning how to do field geology undersea with *Alvin* (and hence these geologists' later resistance to robots' more abstract, data-heavy remote presence in the ocean).

Despite its tacit, intuitive character, field geology became the most prominent science in the Apollo program, with successful collaborations between astronauts and the geologists who trained them. After conquering the early challenge of landing a machine on the moon, the Apollo astronauts turned much of their attention to what they would actually do on the lunar surface. They dived into their geological training, working closely with geologists and planetary scientists to understand how to do fieldwork. Emerging with about the equivalent of master's-degree-level experience in field geology, astronauts took their scientific work seriously. Their innovative traverses of the lunar surface remain unique examples of human activity on other planetary bodies.

NASA recognized this evolution when, for the final flight of Apollo 17, it bumped from the crew an experienced test pilot in favor of a PhD geologist, Harrison Schmitt, who had not yet been to space. It would seem that the inclusion of a scientist on Apollo 17 was a boon for scientists, a confirmation of their importance, and a quantum leap in the quality of the scientific work. Indeed, the National Academy of Sciences and other groups that had lobbied for the crew change applauded Schmitt's presence on the mission.

But not all scientists felt the same way—at least one geologist who was deeply involved thought it was a disaster. He felt Schmitt made some snap judgments on the moon, and preferred the test-pilot astronauts, who, with their masters-level training, behaved like laboratory technicians or advanced graduate students. They served as the disciplined eyes and hands of the scientists on the ground, without trying to make scientific judgments. "A well-trained astronaut can talk just as well as a trained geologist. The other astronauts were better at following the scheme and easier to direct."

To this participant, the astronauts were similar to those performing the Hubble repairs at the end of the robot arm: mobile eyes and hands linked to extended chains of expertise. In fact, Hodges, who still works with Schmitt, describes the Apollo work as "really telerobotics" because "it was a science back room here on earth that was using the astronauts as tools to try and generate data for analysis here on earth."

Jim Head, a geologist at Brown University, trained astronaut Dave Scott for his fieldwork on Apollo 15. Head emphasizes it was also important to turn loose the astronauts to make their own decisions. The critical element of time, with its scarcity amid the ever-present pressure to get things done, meant that humans needed to know enough to move quickly through the landscape (of course the rush was also a consequence of the actual presence of people, who have limited life support resources).

Here it helps to imagine not isolated humans venturing out on the moon, but the astronauts as part of an exploration system. On Apollo they linked by radio to one colleague in lunar orbit, and to the ground in Houston. They rode on a rover, which had a television camera to beam signals live to the ground. On the later missions, in fact, an operator on the ground could control the pan and tilt of the camera and follow the astronauts' activities (it even transmitted live video of the astronauts in the lunar module blasting off for home).

From the remotely controlled camera, it is only one small step to imagine the crews on the ground moving the rover directly through the same interface. Indeed, during the 1970s the Soviet Union operated two lunar rovers, known as Lunokhod, for nearly a year, from joystick-controlled stations on the ground. The low latency of the lunar distances, barely a few seconds, makes this kind of operation particularly rich on the moon. Yet NASA has never sent a telerobotic rover there (though privately funded projects are currently aiming to do just that).

This spectrum of lunar field geology—from scientist explorers to astronaut technicians to quasi-real-time remote rovers—begs the question: what kind of presence is required to explore and do science on other planets?

Let us begin to answer with a thought experiment. Picture the moon walk coverage from the Apollo 17 mission: 22 hours (over 3 sorties), covering about 35 kilometers (22 miles). If that distance traveled were entirely in a circle, it would have enclosed about 100 square kilometers.

Now consider a lunar "hopper" robot. Jeff Hoffman, now retired from the astronaut corps and teaching at MIT, has recently been working on such vehicles. Firing a small rocket engine in pulses can spring these robots on "hops" across the lunar surface. Because of the low lunar gravity—one sixth of earth's—these hops can be kilometers in distance.

In our thought experiment, a lunar mission is launched from earth and lands a pallet containing two hoppers on the moon. Upon landing, the first hopper, specially configured for mapping, unstraps itself and proceeds to hop across an area of 100 square kilometers. High-definition cameras, laser radar, spectrometers, and other sensors build a map of the survey box. It records optical imagery, topography, and other measurements, to super high precision on the order of millimeters. Onboard computers crunch the data before sending it back to earth via a telemetry link where skilled digital mapmakers in Houston assemble it into a millimeter-scale map of the survey box.

These data are then turned over to a team of geologists, who, over the next several months, explore the data set, the topographic map, and the entire survey box through their computer screens. They use virtual-reality goggles or special immersive rooms that simulate driving through the environment. Given the minute scale of the data set, they can virtually "stop" at sites of interest for close inspection (though, of course, they cannot kick over the rocks). Scientists explore together, pausing for in-depth discussions of what they are seeing and where to go next.

In a few months, the team summarizes its findings and identifies sites of interest within the survey area. They come up with a sampling plan and hand it off to a group of engineers who translate it into a set of instructions, plans, and trajectories for hopper number two, which has been waiting dormant on the lunar surface the whole time.

This hopper is designed not for imaging but for drilling, scraping, hammering, and sampling. Given a set of, say, one hundred points, specified with reference to the map built by its companion, hopper number two ventures out, and over the course of several days methodically collects samples and rocks, and returns them to the pallet.

On board the pallet, a suite of laboratory instruments decompose and analyze the samples. An advanced version might even sort the samples, put them in protective covers, and load them aboard a small rocket that takes off from the moon and returns to earth (as the Soviet Luna missions did in the 1970s).

The entire process takes three to six months, much slower than the three days of the Apollo 17 lunar stay.

My point here is neither to propose nor to engineer such a mission, although it is entirely feasible with current technology. Rather, the hopper thought experiment forces the question: What did astronauts do that would elude geologists on the ground exploring superfine-scale 3-D models? Manipulate the soil? Achieve "situational awareness" and "presence"? Interact with the environment in real time? Real experience from the Mars Exploration Rovers helps us answer these questions.

The two mobile robots *Spirit* and *Opportunity* were launched from earth in 2003 and arrived on opposite sides of Mars in 2004. A suite of cameras, instruments, and tools allows them to traverse the landscape for several kilometers, map the area, and drill and analyze rocks. The ultimate scientific goal is to characterize any earlier presence of water on Mars, with the eventual goal of seeking extraterrestrial life.

Though originally designed to operate for only ninety Martian days (known as "sols"), both rovers operated for many times longer. *Spirit* got stuck in the soil in 2009 and lost contact in 2010, while *Opportunity* continues to work a decade after its planned demise. As of 2014 *Opportunity* had driven farther (40.25 km) than the Apollo 17 lunar rover (35.7 km) and farther than the Lunokhod 2 vehicle (39 km), setting a record for off-earth planetary driving.

For all these years the rovers have been directed by NASA's Jet Propulsion Laboratory (JPL) in Pasadena, California. There, engineers and scientists sit in windowless rooms, issuing commands for the rovers, reviewing the data, and generally exploring Mars (after the first few months, the "nominal" missions, many of the scientists returned to their home institutions to participate through the internet).

The rovers are dependent on power for their solar panels, so most operations need to happen during the Martian day, which is about forty minutes longer than an earth day. Sometimes a Martian sol corresponds to the earth day—hence a normal work day for the crews in Pasadena—but other times it is nearly opposite, leading to strange work hours. Some participants began wearing multiple watches to remind them of both Mars time and Earth time. In the long run, the strains on human performance of these unusual schedules, or "planetary jet lag," actually began to pace the rovers' scientific output.

Nonetheless, in the first decade of this century, a small number of

Feeling present on Mars: scientists reviewing maps and plans during the Mars Exploration Rover expedition.

(PHOTOGRAPH BY WILLIAM CLANCEY; REPRINTED BY PERMISSION FROM CLANCEY, *WORKING ON MARS*, PAGE 149)

people learned how to drive to work in the morning and go to work on another planet, much as Predator crews, not far away, were driving to work and going to war on another continent.

It was an unusual kind of work. Geologists who may have chosen their careers because they liked to be outdoors, now found themselves in air-conditioned rooms, looking at screens and going to meetings while still living away from their families. Their work required more collaboration than traditional field geology, coordinating people and machines over time-delayed limited-bandwidth links between two planetary environments.

Bill Clancey is a computer and cognitive scientist who had studied scientists' use of robots while in isolated environments in the Arctic. When he turned his attention to studying workers at JPL to determine how they learned to work on Mars, he became interested in the scientists' experience of presence on the distant planet.

Public and press accounts, including those from NASA itself, often refer to the MERs (Mars Exporation Rovers) as "robotic explorers." But clearly they are not. The robots do not explore on their own, they do not make judgments, and they do not do any science. They are more like remotely operated undersea vehicles than robotic explorers—except with twenty-minute time delays between commands and responses.

Clancey argues the rovers are mechanisms that people "acted through"—extensions of human eyes and hands of people on earth. The rovers are more like programmable, mobile laboratories than scientists, physical more than cognitive surrogates. He writes about the scientists' experience of "becoming the rover." In language analogous to that of Predator pilots, scientists talk about "projecting yourself into the rover," and find themselves turning their heads to look behind the rovers, straining their necks to look around rocks, as though they are physically there. "It's been some kind of weird, man-machine bond," one scientist says. "It has morphed into us and we've morphed into it." Another reports, "My body is always the rover."

Given this projection, what did it mean to call the rovers "robotic explorers"?

This question took Clancey into the worlds of field science and remote presence. He found that the mission reports repeatedly "ascribed machine initiative to remotely controlled actions" such as "*Spirit* collected additional imagery of the right front wheel."

The chief scientist, Steve Squyres, writes compellingly about what he calls "this kind of bizarre combination of planetary exploration, robotics, and management," he did for a living, "in the service of geology on another world." When Squyres writes about his team's work, he often refers to the team itself being on Mars, for example, "As we work across the plains . . ." or "We've arrived at Endurance crater . . .," referring to "the slope immediately in front of us." Much as in the *Jason* control van, Clancey found that the MER team had a deep sense of presence in the landscape they were studying. *"We were all there, together, through a robot!"*

On average, Mars is 140 million miles away from earth, which means it takes about twelve minutes, at the speed of light, for a command from earth to reach the rovers, and about twelve minutes for any results of that action to be perceived on earth (actual travel time varies between three and twenty-two minutes). In practice, this means that engineers on the teams issue commands to the rovers and see the results about once per day.

Observers of these missions often assume that such time delay destroys any possibility for feeling present in the Martian landscape. Clancey's research found exactly the opposite: the delays work into a daily cycle of interactions, "enabling a feeling of synergistic operation, indeed of being there on the planet." With the *Jason* robot scientists learned to turn the control van into a real-time seminar on the ocean floor. On Mars, this daily cycle enabled deep immersion in the data being sent back, actually enhancing the sense of presence.

Spirit and *Opportunity* do not work as autonomous beings, but rather as physical surrogates for the scientists' bodies and senses. Cognitively, the work remains in Pasadena, shifted in space (by millions of miles) and in time (by the daily cycle). As with Predator operators, whose sense of presence heavily depends on their social environment, scientists feel they are working on Mars because their perceptions, their teamwork, the interplanetary system, and the rovers make a kind of cognitive sense. The team on the ground sees things in the world, considers the data and imagery, makes decisions, sends commands to the rovers, and sees the results of their actions, in Clancey's words, "a daily cycle of telerobotic activity, evaluation, and programming." That the whole cycle takes a day instead of the milliseconds that it would take if one were hammering on a rock in an earthly desert is, in the long run, irrelevant.

Now recall Squyres's oft-repeated comments on the slowness of working with rovers: "It took four years to do a week's worth of fieldwork! It has unfolded in excruciatingly slow motion." The sentiment seems odd because it suggests that the major reason to send people to Mars would be *speed*, which few suggest as a rationale. It will cost hundreds of billions of dollars to get humans to Mars for a few months, whereas the rovers have already enabled work to take place on Mars for over ten years for about the cost of a single space shuttle mission.

Clancey interprets Squyres's comments on speed to reflect a measure of the scientists' sense of presence in the landscape. Telerobotics can provide presence, but at some cost—"It distances the scientists from a landscape they would prefer to walk through," Clancey says. The very success of the telerobotic work cycle leaves the scientists wanting more—the rover's affordances of presence are "tolerable but not satisfying," similar to the limits of presence felt by Predator pilots.

Scientists recognize a "fundamental fallacy" in the "Geologists could do in a minute what the MERs do in a day" sentiment. The latency time actually favors analysis of the data, thoughtful consideration, and scientific deliberation among the group before the next move. As in the *Jason* control van, teams of scientists on the ground can talk decisions through before pursuing the next step—new configurations of work in space and time.

One source of the rovers' designation as "robot geologists" was the sense that they act autonomously. But the richer idea of immersive presence in a remote environment mediated by twenty-minute time delays allows us to situate the rovers' autonomy. It makes sense that *Opportunity* should be able to execute local commands during the long delays between earthly commands—and indeed it conducts numerous local feedback loops and housekeeping jobs, controlling the instruments and keeping the rover healthy, with no human intervention.

But in practice, the rover's autonomy serves as a resource for the human engineers who command it. For example, the rover can autonomously plan a route around a series of rocks or obstacles using imagery it gathers from its camera, using a program called AutoNav. But to do that it stops every ten seconds to look at the terrain for twenty seconds. Thus autonomy is costly in time—the rover can drive three times faster when the human planners give it the route in advance. In another autonomous mode the rover can automatically choose rock targets according to criteria set by the scientists. In each case, autonomy provides specific tasks, set up and enabled by the human operators, and there's a trade-off—enabling the rover to arrive more quickly requires many hours of human analysis and planning on the ground.

In Clancey's words, autonomy here is "a relation between people, technology, and a task environment." Once again, autonomy is meaningful within a context. One of MER's robotics engineers was "surprised" that the robot he helped build, when it got into the field, acted more like a "partner" than a free-acting agent, more like a human collaborator than a technical bot. That surprise happens when the engineer's laboratory-bounded notion of autonomy encounters real uses of robots in the world.

Some point out that human presence on Mars would be a more efficient way to do the work. But why the need for efficiency, doing more work in less time? Well, the reply goes, because field time is expensive and difficult to get, you always want to get the most data in the shortest time you can. But with MER, the field season lasted ten years, at least. Moreover, the time between sols was valuably used by the science team to organize their thoughts, achieve consensus, and plan their next moves.

Apollo astronauts often speak of the hurried nature of the work on their missions, having to compress a great deal of work into the time limits imposed by human presence. If science is about intellectual engagement, might not there be value in spreading out the cognition in time?

I have asked field geologists what it is about their work that requires interacting with the environment in real time. What would be lost by slowing down the whole process? After some discussion, Kip Hodges concluded, "I have not thought of a way that field geology demands low latency."

Remember the subject matter: geology. Undersea, on earth, on the moon, and on Mars. These are environments that usually haven't changed for millions or perhaps hundreds of millions of years. There's plenty of time to study them.

Sure, one can think of cases where the real-time dynamics of the phenomena are quicker, and the scientists need to poke and prod in real time. Mud and lava flows perhaps, or the behavior of critters at the deep-sea vents that *Alvin* visits. But for the most part biologists prefer to observe even these subjects without intervention, so simple high-speed recording devices might be appropriate. In planetary exploration, the phenomena under study are *slow*; the difference between a two-week human mission (cost: $100 billion) and a ten-year robotic mission (cost: $1 billion) has no relevance for rocks.

Dan Lester, an astronomer at the University of Texas, argues that we need to rethink our traditional concepts of exploration. Even though the human scientists and their Mars rovers are clearly conducting exploration, he points out, NASA still uses the term "exploration" to refer to human spaceflight, while congressional legislation speaks more generally of "human presence" in space. "When Congress starts using the phrase 'human presence' to authorize a $17B agency," Lester writes, "the phrase takes on some importance."

Why must human presence on Mars require "boots on the ground" when remote presence and telepresence, like those of afforded by *Spirit* and *Opportunity*, can provide a sufficient sense in such a foreign environment?

Lester flags a caveat that, for him, makes space exploration "poorly matched to terrestrial telerobotic pursuits." His caveat is latency—the communications delays for control signals and data. This latency, Lester argues, makes any sense of presence on Mars "downright poor." In his view, presence requires latencies on the order of human reaction time, roughly 200 milliseconds, which are unachievable from Earth-Mars distances.

In 200 milliseconds light travels about 30,000 kilometers (about 18,000 miles), which Lester calls "the cognitive horizon"—closer than that, and we can feel remote presence, whereas farther away we cannot. The moon is six times farther than the cognitive horizon, and Mars is thousands of times farther.

Lester and his NASA colleague Harley Thronson argue that direct human presence in the Martian environment is necessary, if not all the way down to the surface. Going only as far as orbit around Mars requires launching as little as half the total amount of mass from earth as required by the expensive and dangerous forays down onto the surface. Lester and Thronson argue for "on-orbit telerobotics" wherein astronauts in orbit around Mars or some other body control robots down below on the surface. "Exploration derived from human presence may well not need humans in situ," they write, "though it probably needs humans close by. . . . It's simply about relaying human cognition from one venue, which may be relatively inhospitable, to another, which is more hospitable."

The cognitive latency argument laudably moves beyond the old argument that remote presence is not real presence, but in doing so introduces another fallacy: that presence through a time delay is not real presence. Ask Predator pilots about the sense of presence when they operate through delays nearly ten times longer than Lester and Thronson's cognitive horizon. Everything in Bill Clancey's studies of the Mars rovers teams, everything in his richly empirical and systematic data, contradicts this assumption. What is it about latency that destroys presence? Why can we not feel present when our data is a few minutes or even hours old?

When the thing you're studying hasn't changed in millions of years, why is twenty minutes too long to wait? Lester and Thronson accept human presence shifted in space, but reject it when shifted in time.

My goal here is not to argue for or against human spaceflight, the justifications for which have always been, and will continue to be, primarily about engineering demonstrations, national prestige, and international competition more than any cognitive or motor task advantages. Rather, spaceflight offers a dramatic and salient example of the relationships among space, time, task complexity, robotics, and human experience. In low earth orbit, with relatively low latencies, telerobotic systems can accomplish a great deal through direct manipulation. On the moon, with only slightly longer delays, teleoperation offers great potential not yet explored by NASA. Mars, with its much longer delays, requires distributing human action and agency across time, through both work practices and technologies like autonomy, and creating new ways of working. None of this precludes the experience of presence in the Martian landscape, and in fact each enables collaborative presence, new ways of doing science, and new ways of exploring our world and our solar system.

In space, great distances force us to spread out cognition over time, enabling us to see how autonomy maps across the solar system as distributed human presence. How we program our models of the world into autonomous systems here on earth is the subject of the next chapter.

CHAPTER 6

Beyond Utopian Autonomy

"ABE, Pioneering Robotic Undersea Explorer, Is Dead at 16."

NEW YORK TIMES, MARCH 15, 2010

ABE, THE AUTONOMOUS BENTHIC EXPLORER, IS THE ONLY ROBOT to have its own obituary in the *New York Times*. The vehicle, about the size of a small car, was designed to explore the deepest parts of the ocean ("benthic" means the zone around the seafloor).

ABE was lost off the coast of Chile, likely due to an implosion of its pressure housings under the extreme strains of great depth. At the time of its demise, ABE was on its 222nd dive since it first began mapping the seafloor in 1996, and was already in a state of semiretirement, having been superseded by a more advanced vehicle named *Sentry*. While the loss stung ABE's inventors, who monitored it from a nearby oceanographic vessel, in death it confirmed its own value, as no human lives were lost.

ABE was the project I was hired to work on at Woods Hole in 1989. Its original mission was to descend to the seafloor near a series of hydro-thermal vents, moor itself to some kind of latching device, and go to sleep. The plan was that it would then periodically wake up—perhaps once a day for a month, or once a month for a year—and conduct a pre-cise survey of the vent area, making measurements, taking pictures, and documenting the growth and decay of the unusual geology and ecosystems. After designing some of ABE's early computers, I moved on to other projects, while ABE's principals, Dana Yoerger, Barrie Walden, and Al Bradley carefully nurtured ABE through a long matu-ration.

ABE never conducted its original mission. What it did do, however, was map the seafloor with pioneering precision, which meant running straight track lines back and forth across wide areas while collecting massive amounts of data on the topography. "ABE was never meant to fly in a straight line," Yoerger recalls, "but you go where the business is." Scientists were willing to pay their precious grant money for geologic-scale maps, so that's what the ABE team learned how to do.

ABE evolved in concert with the manned submersible *Alvin*, not as a replacement for it. Barrie Walden, one of ABE's three principal engi-neers, was the head of the *Alvin* group at Woods Hole. Most of ABE's early dives were conducted on *Alvin* cruises, from *Alvin*'s mother ship at night, while the sub was on deck charging its batteries. *Alvin* even came to ABE's rescue when it failed to return from one of its early dives.

In 1999, on a cruise to the East Pacific Rise, two days' sail from Easter Island, it all came together. Yoerger had planned a series of mea-surements to gather magnetic data across a ridge. But geologist Bill

Ryan, originally trained as an engineer, pushed Yoerger to program ABE to do more systematic surveys than local measurements. ABE went down 2,600 meters (nearly 8,500 feet), followed about 20 meters (65 feet) above the rough volcanic terrain, collected sonar and photographic data, and returned to the surface before dawn. Yoerger found that if he worked quickly he could offload the data, create a first-cut map, and print it out in time to hand to the scientists about to dive in *Alvin* in the morning.

With a little more time, Yoerger assembled track lines from eight individual dives into a single map covering an area about 1x4 kilometers. Yoerger remembers handing the first map he made, with some trepidation, to the Icelandic geologist Karl Grönvold, and asking, "What do you think?" He was met with silence. "Just like almost all geologists," Yoerger recalls, "when you hand them a map, you know you've got them when they don't say anything for a couple of minutes." Grönvold was parsing for consistency, patterns, geological details. Then he looked up at Yoerger and said, "I don't have maps this good of Iceland." Autonomy was proving its value by putting data in the hands of human scientists.

For the first time, scientists descending in *Alvin* could enter the rugged volcanic landscapes with a real map to guide their explorations. In retrospect, it seems absurd that some vent fields had been visited by hundreds of *Alvin* dives, but nobody had yet mapped the areas. The autonomous vehicle brought the quantitative side of geological mapping to *Alvin*'s immersive field-geology methods.

After this breakthrough, ABE went on to more than a decade of geological mapping, constantly improving in navigation precision, the density of the data, and the quality of the imagery. Its temperature and chemical sensors could even detect the rising plumes from hydrothermal vents, and by 2004 Yoerger and his team developed methods to employ that data to direct ABE to discover new vent fields.

The Autonomous Benthic Explorer (ABE) running survey track lines over a hydro-thermal vent field, using a scanning sonar to collect bathymetric data for precision mapping. ABE's measurements of temperature and chemical data are represented by the shaded line behind the vehicle, illustrating its detection of a hydrothermal plume.

(COURTESY CENTER FOR ENVIRONMENTAL VISUALIZATION, UNIVERSITY OF WASHINGTON)

ABE's successor, *Sentry*, was designed specifically for this new mission. In 2010 it used similar techniques to map the underwater plume of oil emanating from the Macondo well in the Deepwater Horizon oil spill, and showed that much of the spilled oil was not washing up on shore but was suspended undersea in an enormous cloud.

What constituted ABE's autonomy? The absence of a tether certainly enabled ABE to operate freely, and to get in close to rough terrain in ways that would be difficult with an ROV like *Jason*, or dangerous with *Alvin*. A great deal of clever design and programming was required simply to make the vehicle safely go out and come home again without hitting anything, with enough smarts to surface and call for help in case of problems. The basic paths, though, were simple, straight track lines, back and forth, to cover the maximum terrain without missing anything. As in our lunar hopper thought experiment, the exploration was still carried out by the scientists as they read through the data, interpreted the maps, and became present in the landscapes.

What's more, ABE's dives actually became less autonomous over time as the technology for acoustic communications improved. The early ABE dives were operated largely independent of human input—maybe a few acoustic pings to signify that the vehicle was still alive, perhaps the ability from the surface to send a single acoustic code to abort the mission. "Other than that and looking at the topside [surface] tracking, we had no idea what was happening below," engineer Rich Camilli recalled.

Gradually, however, the ability to transmit data through the water improved. Underwater acoustic modems can send data packets a few kilometers through the water, like the old telephone modems that translated computer data into a series of awful-sounding bleeps and burbles, and about as fast. With these modems, AUVs like ABE could upload data packets, like text messages, to report on vehicle status, navigation, depth, battery life, and even scientific sensor data and images. At the same time, humans on the surface could download new commands to the vehicle, potentially redirecting the survey.

On one ABE dive in 2009, the main onboard navigation gyros failed. To continue the valuable foray, the engineers on the surface gave ABE simple commands to move a few meters at a time in this direction or that, essentially "joysticking" the vehicle through the slow link (much as Russian engineers had joysticked the Lunokhod rovers on the moon in the 1970s). The problem then became not increasing the autonomy of the system but designing displays and algorithms to help the humans analyze the data in real time.

Like the communications to Mars, acoustic communications under-water have limits. The bandwidth is a fraction of even the most ordinary WiFi network in a coffee shop, and the time delay over long distances measures on the order of seconds—as long or longer than sending radio transmissions to the moon. ABE's or *Sentry*'s operators, while they like to stay in touch with the vehicles when they can, will still move the ship far away to do other jobs during a long dive. During those periods the vehicle is more autonomous.

Even were the vehicles out of touch for an entire dive, they still regularly return to the mother ship. Engineers tend to think of the autonomy beginning when the vehicle dives. But in a broader view, the exploration system is a manned vehicle (the ship) launching the unmanned vehicle for periods of autonomy, until it returns to its human operators to exchange data and energy and receive instructions. As on the Air France 447 search, autonomy is periodic, mediated by regular human interaction and dependent on position, bandwidth, and a host of other factors. Yet again, autonomy exists within a context.

James Kinsey, a young engineering scientist at the Deep Submergence Lab, came to the job with great plans for the autonomy he hoped to bestow on the vehicles. He began to build up probabilistic models of how the hydrothermal vent plumes propagate through the ocean, and to try to instruct the vehicles to follow minute detections from their sensors down to the vents. Over time, however, Kinsey realized that "trying to imbue that much autonomy into the vehicle was likely to be a problem from the beginning." Because of the nature of oceanographic exploration, the tasks are poorly defined and the environment is changing. Anything programmed into the vehicle constituted assumptions, models about how the world worked that might not be valid in a new context. "I think it focused on the wrong aspects of autonomy, perhaps. . . . You're requiring the vehicle to understand a lot of context that may not be available to it." Kinsey had his own version of the surprise expressed by the Mars rover engineer as abstractions of autonomy met real applications.

"One of the problems with having a vehicle that makes its own decisions," Kinsey realized, "is there's a certain amount of opaqueness to what it's doing. Even if you're monitoring it, [you say] 'Gee, it just suddenly wandered off to the southwest. Is it malfunctioning or is that part of its decision making tree?'" Operating in the deep ocean is expensive, and autonomous vehicles, even though they're unmanned, are far from disposable. "People like to know where their assets are," Kinsey observes, "especially when they pay a lot of money for them."

Moreover, communications technology is not static. Optical communications, for example, are becoming practical. These basically consist of LEDs that flash at high speed to send data through the water; they can achieve WiFi-like rates over short distances, on the order of a hundred meters. With this technology, a ship might dangle an optical modem down to the depths like a streetlight. Then an AUV might divert its route to make a communications pass by the light, upload its data quickly, and then be on its way. Or a person on the surface might even teleoperate the vehicle when it's in optical range, then let it do more on its own when out of range or if the optical link is lost. Autonomy then becomes a function of position and bandwidth.

Overall, the lines between the human, remote, and autonomous vehicles undersea are blurring. Engineers now envision an ocean with many vehicles working in concert. Some may contain people, others will be remote or autonomous, all will be capable of shifting modes at different times. The recently upgraded *Alvin* has software originally designed for autonomous vehicles; one day it may connect to the surface with an optical fiber. One day it might even operate unmanned.

The challenges are to coordinate all of these machines, keep the humans informed, and ensure the robots' actions reflect human intentions. Some will operate through high-bandwidth channels like optical fibers, others through more constrained channels. Some will circle close to a node to flash their data, then slide back into the abyss. Each will do as it's told and make some decisions on its own, according to structures encoded by its human programmers.

In this emerging world, we can imagine autonomy as a strangely shaped three-dimensional cloud in the ocean, with vehicles constantly moving back and forth across its boundaries. Now imagine that ABE is your car, and the 3-D cloud of autonomy is in your neighborhood. At certain times, in certain places, the car has some kinds of autonomy—to stay within a highway lane, for example, or drive in a high-speed convoy. At other times, such as when far from a cell tower, or driving in snow when ice obscures the car's sensors, the autonomous capabilities are reduced, and the driver must be more involved. You drive into and out of the cloud, delicately switching in and out of automatic modes.

Indeed the prospect of driverless cars—they might have been called "automobiles" had that term not been taken a century ago—is generating broad public enthusiasm and debate around autonomy. Google has been the most outspoken player, with a carefully controlled publicity campaign. (Most of their work has been proprietary so we must rely on public statements to assess the program.)

Automobile manufacturers, of course, have been adding various kinds of automation to their cars for decades, from automatic transmissions to cruise control and antilock brakes. My family's Volvo has software that can slam on the brakes at any moment if it detects the car is about to hit something. (When driving that car I must place a great deal of trust in that software.) In general, auto manufacturers are continuing their incremental approach, selling automation as "safety features" rather than autonomy. Mercedes-Benz has announced a concept car with an interior envisioned as a "mobile living space," where people can read and relax under the direction of the car's autonomy. Unlike the Google car, however, Mercedes imagines that "passengers are able to interact intuitively with the connected vehicle" in a "symbiosis of the virtual and real world." Google, on the other hand, has been promoting a vision of complete autonomy. As one Google engineer compares their approach to those of the car companies: "They want to make cars that make drivers better. We want to make cars that are better than drivers." The ride-sharing giant Uber recently hired a large group of roboticists away from Carnegie Mellon, in an apparent effort to automate their cars.

Google has been testing self-driving cars on California roads since 2009, claiming hundreds of thousands of miles of accident-free highway driving. They travel routes mapped with great precision by Google's human-driven survey cars; the maps serve as virtual railway tracks for the cars (indeed, they are as yet unable to drive on roads without these detailed maps). The drives have included human safety drivers and software experts who can turn the autonomy on and off. "The idea was that the human drives onto the freeway, engages the system, [and] it takes them on the bulk of the trip—the boring part—and then they reengage," said Google engineer Nathaniel Fairfield.

A ride in one of these vehicles led the *New York Times*'s John Markoff to conclude that "computerized systems that replace human drivers are now largely workable and could greatly limit human error," potentially supporting Google's goal of cutting the number of U.S. highway deaths in half. Google's rhetoric around the project has the kind of Silicon Valley optimism that typically surrounds software systems. Roboticist

Sebastian Thrun, lead engineer for the project, envisions a future of utopian autonomy "without traffic accidents or congestion."

A number of critics of Google's approach have pointed out its limitations. Most of the work has been done in northern California or other western states. The Google car's successful driving tests in Nevada were run under tight constraints from the company for good weather and simple routes (the company also sought to avoid disclosure of details of the safety drivers' disengagements of the autonomous operation). The vehicle's algorithms had difficulty negotiating construction sites, requiring the safety driver to take control. Technology journalist Mark Harris has recently shown that becoming a safety driver for one of these cars can require weeks of training, suggesting the computer to human handoffs remain complex and risky.

In contrast to the wide-open West of car commercials and Google's trials, urban driving entails a great deal of social interaction, as we drive through a messy, complicated, and dynamic physical and social landscape. Google admits this problem is ten to one hundred times harder than driving on highways. Once again, autonomy within a human context proves much tougher than the abstracted technical problem.

MIT's John Leonard, who helped develop some of the basic algorithms that driverless cars use for localization and mapping, points out how much driving depends on social interaction. My late friend Seth Teller, formerly an MIT roboticist, perceptively observed that urban driving consists of hundreds of "short-lived social contracts between people," as we scan the streets, make eye contact, let people in and wave "thank you." Computers are slowly getting better at assigning labels to the physical world and to different kinds of objects. Yet as Predator pilots can attest, techniques to similarly interpret human identities and intentions remain primitive.

Only half joking, Leonard contends that driving in Boston can be considered operating in an extreme environment. He put a video camera on the dashboard of his car and is collecting examples of driving situations that are difficult for algorithms to handle: merging onto a busy road at rush hour; staying within road lines obscured by dust or snow; turning left across several lanes of traffic. In the snowy Boston winter of 2015, the three-dimensional landscape of urban driving would change overnight, as snow piles nine feet high narrowed the roads and altered traffic patterns.

What have we learned from extreme environments that might shed light on possible futures for autonomous cars? We know that driverless cars will be susceptible to all of the problems that surround people's use of automation in the environments we have examined—system failures, variability of skills among users, problems of attention management, the degradation of manual skills, and rising automation bias as people come to rely on automated systems.

The most challenging problem for a driverless car will be the transfer of control between automation and the driver—what we might call "the Air France 447 problem." Any life-critical system has to have ways to handle anomalies when an individual sensor or component fails, or when things in the world just don't work out as expected. The more complex the system, the more potential anomalies hidden in the corners. While these anomalies may be rare, there are more than a billion car trips per day in the United States, greater than ten thousand times the number of daily airline flights.

Google's car might recognize a situation it couldn't handle and warn the driver to take back control. Perhaps it will have a "check autonomy" light, analogous to the opaque "check engine" light in your car today—though it will have to be much more informative to be effective. What happens when the light goes on at high speed and the trade-off has to happen quickly? If people are freed from the tedium of driving and are sleeping, or reading a book, how will they rush into the loop quickly enough to avoid oncoming traffic, or a pothole, or a collision?

As on Air France 447, drivers would have to rise to this challenge at the most difficult moments of driving. And in contrast to the operators we have been following in extreme environments, automobile drivers are relatively untrained and comprise a broad spectrum of physical abilities, acquired skill, and socioeconomic and educational backgrounds. As challenging as they are, the extreme environments of the deep ocean, air, and outer space are relatively "clean"; the driving environment is much more cluttered and uncertain.

Perhaps in response to these critiques, Google changed its approach to get rid of these troublesome people—drivers. As engineer Nathaniel Fairfield says, Google discovered that "people are lazy," and found that they "go from plausible suspicion to way overconfidence." Following its experiments, the company concluded that human drivers are not trustworthy enough to collaborate with their software and changed their model to zero driver intervention. Google engineers speak about the "lazy driver," the 93 percent of car accidents estimated to derive from human error. (Of course, human-factors specialists have long understood that human errors often are the result of poor system design and poor work practices.)

Google introduced a new version of its car in 2014, one that seems designed to be friendly and unthreatening to the public. It travels at low speeds to reduce danger, has no driving wheel or console, and removes input from the human driver altogether. In the words of project director Chris Urmson, the company is "working toward the goal of vehicles that can shoulder the entire burden of driving." These fully autonomous cars would be "designed to operate safely and autonomously without requiring human intervention."

The new car's interface consists only of buttons to start and stop the engine, and a screen that shows the route (one wonders how the driver will tell the car where to go). The driver will be transformed into a passenger whose only job "is to kick back, relax, and enjoy the ride." "It doesn't have a fallback to human—it has redundant systems," said Fairfield. "It has two steering motors, and we have various ways we can bring it to a stop." Videos set to lighthearted music appeared on YouTube showing blind, disabled, and elderly passengers enjoying pleasant rides in the California sunshine on clean, open roads.

Providing mobility for the disabled or elderly is certainly a laudable goal, but Google's new approach amounts to a retreat and a missed opportunity. Ironically for such a high-technology company, Google rhetoric takes a step backward into the twentieth century, archaically reimagining the driver as a passive observer. Their "new" approach succumbs to all three of the twentieth-century mythologies of robotics and automation: 1) automobile technology will logically advance toward complete, utopian autonomy (the myth of linear progress); 2) autonomous driving systems will eliminate human drivers in the driving task (the myth of replacement); and 3) autonomous cars can operate entirely on their own (the myth of full autonomy).

Our ventures into extreme environments have taught us how this utopian autonomy succumbs to the realities of harsh environments where lives are at risk. Indeed Google's utopian autonomy is a more brittle, less functional solution than a rich, human-centered automation. It's true that including human drivers in automated cars raises problems of mode confusion, attentiveness, and interfaces, but these are tractable— especially using what we have learned from extreme environments.

Instead, Google engineers are succumbing to naïve automation, defining the problem in ways that Google-enabled software can solve. This definition appeals to engineers because of its self-contained nature: driving as purely a problem of navigation and collision avoidance.

This definition may be acceptable for narrow, local applications (I would not be surprised to see such driverless cars in niche applications, ferrying passengers in parking lots or college campuses, much like the monorails at airports do). But the full spectrum of driving as a social activity is remarkably broad, encompassing a diversity of geographic, economic, cultural, and other components. Google presents no evidence that their code even recognizes this wonderful variation. Moreover, simply having people in the car raises ethical stakes of risk, agency, and reliability that have to be addressed. Google's engineers repeatedly congratulate themselves for rising to the most difficult challenges, yet here they have punted the meaningful, socially beneficial problems in favor of narrow algorithmic solutions.

And, of course, even with Google's autonomous cars, the people are still "in there," they've just moved to a different place and time. Let's look inside an algorithm as an example of how deeply humanly crafted apparently autonomous code can be. Consider the first documented collision between autonomous cars. The Defense Advanced Research Projects Agency–sponsored competition, the DARPA Grand Challenge of 2007, generated some of the technology on which the Google car is based. Google's Chris Urmson was the lead engineer on the winning team, and a number of other participants are now on the Google team.

In the incident, the MIT car, called *Talos*, was passing the Cornell car, dubbed *Skynet*, which was having trouble with its planning algorithm and was stuttering along slowly by the side of the road. Computers on board *Talos* classified *Skynet* as "a cluster of static objects" rather than a moving vehicle, and turned to pass in front of it. But the Cornell car was not stationary; it was moving in a lurching pattern that *Talos* did not recognize. *Skynet* bolted forward just as *Talos* cut in front of it, leading to a small collision. Neither team won the competition.

To their credit, the teams got together and published the details of the crash. Numerous algorithms and sensors were involved, but a key element was the MIT car's failure to classify the Cornell car as a moving object and to derive its future path. Ironically, the MIT strategy was to avoid classifying objects as one thing or another ("car" or "guardrail"), which can be prone to errors, and instead to only classify them as moving or not moving. But when analyzing objects detected by the vehicle's sensors, the velocity data contained random noise (as all data does), so the MIT system used a threshold velocity of 3 meters/second to filter the data. Anything faster than that was considered "moving" and anything slower than that was considered "not moving."

How was that threshold set? By one engineer estimating the difference between stopped and moving and dialing it in to the algorithm. I asked my colleague Jon How, one of the principals on the project, how many such thresholds there are in a system like that. His reply: "Many, many, many." In fact the "configuration file" for the MIT vehicle contained nearly a thousand lines of text, setting hundreds of variables: sensor positions and calibrations, fudge factors to align the sensors with one another, how to deal with sun dazzle, etc. Machine learning techniques can help reduce this reliance on parameters, but they still rely on human programmers for their basic structure. How points out that core algorithms generally rely heavily on accurate models of uncertainty in the world. As he observes, "The problem of autonomy is fundamentally the problem of living in an uncertain world."

This brief look inside an early autonomous car's code points to how deeply such "autonomous" cars are suffused with human judgment, in countless little details like the threshold we looked at, and in more profound ways, like their models of uncertainty. Recall our original picture of autonomy as a set of pipes that takes in sensory inputs and transforms them into goal-directed actions. It's wonderful technology to behold, but the pipes and transformations are human designed.

In a vehicle like a car they can kill you.

Lawyers and legal scholars are just beginning to consider the liability problems involved in driverless cars. If your notion of autonomy is that the vehicle is making decisions on its own, then the chain of intention that constitutes liability might be broken. Who is responsible when your Google car drives you into a ditch? This is not just a matter of letting lawyers write contracts, but gets to the fundamental notion of autonomy: if the system is really working on its own, then how can it be the manufacturer's fault when something goes wrong? (Some believe that traditional product liability will apply here just fine: if the company makes the product, they will be liable). More practically, how does one certify as safe the software in an autonomous car?

The certification approach for software on life-critical systems like airliners is fairly robust, but cumbersome and expensive: rigorous testing, running through every possible piece of code at least once, careful control of changes. These standards also certify the human processes of planning, designing, and writing the code according to rigorous requirements, as well as quality assurance and managing upgrades once it's released. But in their current form these procedures are not compatible with systems that claim full autonomy, where the number of possible courses of action verges on the infinite. What's more, like synthetic vision systems in aviation, autonomous cars like Google's must rely on high-integrity databases served with frequent updates. Miss last week's update, and you could drive right into a construction site, or a snow pile.

How will we certify Google's models of uncertainty and risk? Every automated path-planning algorithm contains some version of these unknowns. The planning works by optimizing "cost functions"—constantly asking, What is the least "costly" (in terms of time, energy, risk or some other variable) way to get from here to there? But the cost functions themselves embody human judgments about priorities. On one drive, with your kids in the backseat, you might like to drive conservatively; the cost function here should weigh safety more highly than speed. On another trip, perhaps you're alone and in a hurry, and you'd like to push the performance a bit more, take on higher risk. Perhaps you're running low on fuel and would like to raise the importance of fuel efficiency.

As a thought experiment, consider whether your autonomous car should have a knob on it labeled "risk." Want to get home faster? Dial up the risk knob. The system drives more aggressively, you get home a little sooner, and an additional insurance premium is automatically debited from your account. (And what about the other drivers you're putting at risk? Should you contribute to their insurance too?) Driving with your kids in the back? Dial down the risk knob and follow the traffic laws to the letter.

Consciously or not, we make these decisions every time we get behind the wheel. In Google's automated car, an engineer in a cubicle somewhere is making those decisions for you. If we'd like to make these decisions ourselves, then we'd need an interface. What does the map of your neighborhood look like when it's highlighted for risk? When the 3-D clouds of autonomy are made visible?

These thought experiments lead us to an alternative approach to Google's—rethinking, rather than eliminating, the driver. The enticing, shadowy pictures of the laser-scanned landscape that reveal the autonomous car's internal models now become the basis of a new interface, and a new experience of driving. As my colleague Bill Mitchell used to say, "The dashboard should be an interface to the city, not an interface to the engine."

Gone are the tense, shoulder-tightening feedback loops of lane-keeping and speed control. Now we are in a supervisory role, one that commands high-level vehicle behaviors, but still allows time on the wheel. We use rich, sensor- and algorithm-enhanced models of the

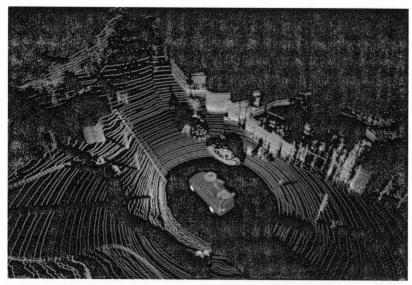

Screen capture image of the representations inside an early driverless car. The circles emanating from the car represent the scans of a laser radar (LIDAR), which detects the road and trees alongside it. The algorithms then filter and classify these data to generate paths for the car to follow. Could this type of image be the interface for a new way of driving?

(COURTESY EDWARD OLSON, MIT)

environment to move across a spectrum of automation moment by moment, driving into and out of clouds of autonomy and risk.

It will not be easy to get this right, but solving this problem has greater potential than putting our hopes into a utopia of full autonomy. We should be able to aid the elderly and disabled, enable other tasks (like texting or reading) while driving, and enhance safety, while still keeping central the value of human presence.

Google's goals and rhetoric for their autonomous cars have changed regularly, and are likely to change again in a fast-moving public conversation. Their public statements likely reflect some difference of vision among the members of their engineering team. Yet the company's cultural power is such that they have framed the debate, from state legislatures to car dealerships, and they are not alone. Both the National Highway Traffic and Safety Administration and the Society of Automotive Engineers define levels of automation in cars that explicitly or implicitly embed the myth of linear progress in a progressive series that culminates in "full automation." Neither group's standards explicitly allow for hybrid modes where some tasks and not others might be highly automated, nor for information-rich automation that might keep the driver involved.

What the Google car proponents have missed, and many in the technology press have also overlooked, is that the liability, certification, and risk issues are not incidental "societal" issues that stand in the way of this otherwise independent invention. Rather, they are crucial parts of the question of autonomy, and crucially involved in framing the future of our robots and ourselves. It is not just a technical issue; it is about who is in control.

As examples of this alternative approach, I conclude with two projects in their early stages that seek to enhance users' roles and awareness

of the internal states of autonomy. Each seeks to engineer a human/
machine team from the very beginning, rather than to design a highly
automated machine to which a user must adapt.

An unmanned helicopter approaches a landing zone at high speed. It
scans the terrain with a laser—much like the laser scanner on top of
the Google car—that zips across the surface and takes thousands of
measurements. The laser feeds a digital 3-D model of the topography
being built inside the computer in real time. Inside, simulations fly flight
paths through the model into the immediate future, applying sophisticated algorithms to the data. They identify where the terrain flattens
out and determine where there is enough room, free of trees, wires, and

Autonomous helicopter of the Autonomous Aerial Cargo/Utility System program
(AACUS) scanning, selecting, and approaching a landing zone in a mountainous
area to deliver cargo or perform medical evacuation. The landing support specialists
on the ground negotiate the landing zones with the vehicle through an iPad interface and a radio link.

(COURTESY AURORA FLIGHT SCIENCES)

obstacles for the helicopter to land. They analyze whether each area is flat enough to ensure the helicopter will not roll over when it touches down.

As the helicopter approaches the landing zone, trees block its view. Relying on the laser scanner, the computer finds a path a little to the right where there is a gap between the trees, commands a line through it, and the vehicle lands.

This scenario describes a demonstration from a real project, a full-size autonomous cargo helicopter built by Aurora Flight Sciences, a Manassas, Virginia–based maker of unmanned aircraft, as part of a government research program. It was demonstrated in flight in February 2014. The program's goal is to send unmanned helicopters into hazardous areas to deliver cargo, and potentially evacuate casualties, with no risk to pilots. I was part of this project team and designed its architecture for human interaction with autonomy. The idea was to consider the autonomous system as part of a human/machine team, not only when designing the interface, but when designing the core algorithms too.

After the demo, the *Wall Street Journal* reported on the project with the headline: "Navy Drones with a Mind of Their Own." In the press, it seems, the myth of full autonomy is alive and well.

Yet where were the people? Everywhere, it turns out. Because this was a prototype, a safety pilot sat on board, his arms folded, watching the computer fly but ready to take over and kick it out of the loop at a moment's notice. A mature fielded system would ideally eliminate the safety pilot altogether. But the safety pilot may turn out to be not as redundant as we think—he or she may just be doing a new kind of job.

More important, when the helicopter lands, a person waits in the landing zone. After all, what good is a cargo delivery without someone, or a group of people, to unload, unpack, and consume the stuff? Those people will need to have courage and a great deal of trust in the machine, for standing in a field with a large helicopter under software control bearing down on you is not a relaxing place to be. Our team interviewed dozens of people, known as landing zone specialists, who do this for a living with piloted helicopters. Most had experience in Iraq and Afghanistan, and a number recalled looking up into the sky and seeing unmanned vehicles flying around. They had the uncomfortable feeling of not knowing to whom the vehicles belonged, or what missions they were performing. The last thing these people wanted working with them in a war zone was a "drone with a mind of its own." What they wanted was a reliable partner that would do as it was told.

This meant the people in the landing zone had to be able to reject the landing. So we gave the landing zone specialist an iPad. With a few minutes of training he or she could interact with the helicopter in a brief negotiation. The human would suggest a landing area; the computer might reject it if the zone did not meet its safety constraints. The computer would then offer a few alternatives. The human could either pick an alternative or command the helicopter to wave off.

As it turned out, designing this negotiation, an interface and software system to execute it in under a minute's time, and a set of internal states for the autonomy that would be comprehensible to the human, proved among the most challenging parts of the program.

Yet the helicopter's successful demonstration raised a question for human pilots: if we can add a scanner and algorithms good enough to identify landing zones, wouldn't a human pilot want that too? As the program moves forward, it may well involve adding these autonomous capabilities to a piloted helicopter as well as to an unmanned one. Medical evacuation helicopters in our own communities, for example, land in uncertain environments under extremely demanding conditions. We are beginning to explore how these novel sensors and algorithms may help their human pilots achieve higher levels of performance and safety.

Some of these ideas are already coalescing into the new notion of optionally piloted aircraft, or OPAs. *Aviation Week & Space Technology*, the industry's leading magazine, has been publishing its Pilot Reports on new aircraft for decades. Its 2012 pilot report on an aircraft called Centaur was the first in which the pilot doing the test never touched the controls.

The reporter conducted the test while sitting in the backseat of the small, twin-engine aircraft. Up front sat a person acting as the safety pilot, his arms calmly resting on his lap. Resting beside him, in what is ordinarily the copilot's seat, was an engineered series of linkages, actuators, and servos. The safety pilot pulled a lever to engage the mechanisms, and

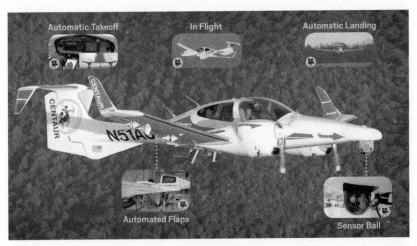

The Centaur optionally piloted aircraft (OPA), a highly modified twin-engine general aviation aircraft built by Aurora Flight Sciences. The Centaur can be flown by a pilot in the front seat, remotely from a computer interface on the ground, or from the same computer interface from the backseat.

(COURTESY AURORA FLIGHT SCIENCES)

they began moving the pilot's control stick and pressing the rudder pedals. The actuators are double and redundant; if one set fails, another will immediately take over. The safety pilot can disengage the mechanism with a single pull of the lever if something goes wrong; otherwise he does not touch the controls.

In the backseat, the "operator" commands the plane through a laptop, using an interface identical to that of the ground control station for an unmanned vehicle. Through the screen, he can change altitude, fly to waypoints, take off, or land. Pushing the "launch" button begins an autonomous takeoff. The computer holds the brakes, pushes the throttles forward, checks the engines and instruments, and releases the brakes for the takeoff roll. The plane accelerates, takes to the air, and begins to climb out on a semi-autonomous flight.

As an OPA, Centaur—named after the mythical half-human, half-horse creature—can fly in a normal mode under pilot control, a certified aircraft as though fresh from the factory. But it also can be flown in semiautomated mode, as in the test when the reporter issued high-level commands through the laptop in the backseat. Or it can be flown unmanned, with the same computer interface operating on the ground through a radio link.

Engineers at Aurora made Centaur by converting a commercially available aircraft, the Diamond DA-42, to this OPA. They added the mechanism in the front seat and a host of independent electronics, and digitized the flight manuals and emergency procedures into computer algorithms.

Unmanned aircraft are currently mostly illegal in U.S. national airspace, but Centaur is certified by the FAA to fly in its unmanned mode if a safety pilot is aboard. Hence the aircraft can help develop sensors, algorithms, and procedures for unmanned systems. It has been flown with a pilot aboard to Alaska, where it has permits to operate unmanned to collect data for climate research.

Centaur employs a transitional technology, suitable for flight test-ing and engineering development while regulations and techniques are worked out. Within it, however, are the seeds of a new way of piloting, for eventually, even in U.S. airspace, the safety pilot's job may be trans-formed to one operating through the laptop interface.

Are we moving into a future of airliners without pilots? Probably not for the foreseeable future, but Centaur shows us how much of the technology that makes that possible exists today. In a technical sense, the automation to taxi an airliner, command a takeoff, follow a route, and autoland is all well proven. The unmanned-aircraft problem appears to have been solved—but only when divorced from its human context. Again, we have learned that fully autonomous operation is the lesser problem. We can say that the technology exists today, but for the inno-vation to have a social importance and contribute to human welfare requires not only the machines but the social, political, and economic systems to situate them in human life and to enhance our practices and our experiences—a much more open problem.

Airliners need to be safety certified so that they don't fall on people. They need to provide not only statistical safety for passengers, but also the experience of safety. They need to operate not only in every emer-gency we can think of, but in nearly all of the emergencies we *can't* think of. This is why, when placed within human settings of reliability, risk, liability, and trust, the unmanned aircraft problem, like the driverless car problem, has not been solved. Building trust in such systems will require years of demonstrations, operations, and smart engineering to prove reliability.

More likely, and certainly sooner, we will begin to see airliners operating with reduced crews. Long-haul flights routinely bring three or more pilots along, to trade off duties even during the dull hours en route. Reducing the number of people required will have a direct impact on costs, if it can be done at an equivalent level of safety.

The trouble is, large airliners are not certified for single-pilot operation. But as expensive pieces of capital equipment they will be with us in their current forms for decades to come. So to reduce the crews we'd have to build add-in devices that can interact with the airplane as a human does now, and interact with the remaining pilot in an effective way. Were we to replace a copilot with a machine during a long flight, the machine might only need to monitor for emergencies and handle the problems only long enough to wake up a napping pilot and make him or her aware of the situation. Such a device, of course, would have to pass the "Air France 447" test of a sudden, challenging alarm—but people have not always passed that test either. (The recent Germanwings crash highlights the challenging social problem of how to safeguard against suicidal pilots.)

The Defense Advanced Research Projects Agency, which also sponsored the early driverless car competitions mentioned earlier, has laid down the challenge of how to add automation to any aircraft quickly and cheaply. If one were to build such a technology, it might support a pilot in a variety of other ways as well—performing routine tasks, looking up procedures, drawing on the database of past events, offering suggestions to improve performance. Avoiding the myth of replacement, it would not do the copilot's job, but would rather depend on a new division of labor between the pilot and the assistant, allowing each to play to their strengths. Thinking this problem through, and building a prototype, is part of a research program sponsored by DARPA; the result may be a system that can serve in this copilot role, but it could also turn any aircraft into an unmanned or remotely piloted vehicle.

The idea is to build some kind of device that will sit in the copilot's seat and interact with the cockpit as a human does—by reading the instruments on the panel with machine vision, grabbing hold of the flight controls to "fly" the aircraft, and even flipping switches and grabbing levers around the cockpit. The project is called the Automated Labor In-cockpit Automation System (ALIAS).

ALIAS has multiple, overlapping goals. Probably the most far out is to turn any aircraft—including a large transport aircraft or a helicopter—into an AUV, with minimal invasiveness to the existing certified aircraft. With the right data link, an ALIAS-enabled aircraft could be flown remotely. A more immediate goal is for ALIAS to serve as a "pilot's assistant," helping with routine tasks during high-workload periods (much as the small robot R2-D2 assisted Luke Skywalker in *Star Wars*). Then the human pilot could operate the entire flight by commanding tasks on an iPad-like interface from the front seat.

The pilot's assistant could enable flying long-duration flights with reduced crews. Rather than taking the place of the copilot, ALIAS entails rethinking the relationship between pilot and copilot—rethinking the basic task of flying an airplane. Long before we see fully automated airliners, we will see flight enabled by ALIAS-like solutions, likely at first on long-duration cargo flights over water.

I am part of Aurora's team that won a contract to design and build ALIAS. ALIAS will use machine-learning techniques to adapt to a new aircraft, machine vision to see and interpret cockpit displays, and some form of robotic actuators to move the controls and flip the switches. It will need training to adapt to new aircraft, which includes the basic flight manuals and procedures but also involves "watching" a human pilot fly the aircraft and gathering data on how the person performs. It will seek to encode the techniques of the most qualified pilots.

The major challenge of ALIAS, however, is not the algorithms but the cooperation with human partners. As we've learned from commercial airliners, the last thing any cockpit needs is another black box with a complex interface and unclear internal states to confuse a pilot. The challenge, then, is to rethink the pilots' tasks in such a way that raises the level of abstraction they work at without adding undue complexity, which of course requires rethinking the role of the pilot.

A pilot may fly in a manned mode one day, in a remote mode the next day, and autonomously on still another. Moreover, the human pilot may choose to let ALIAS fly for one hour and intervene manually in the next. ALIAS-like systems might even allow the pilot to sleep on long flights, provided one can demonstrate the machine's ability to handle the aircraft in an emergency for long enough to wake up the pilot, inform him or her of the situation, and enable him or her to intervene as necessary. Allowing the pilot to nap is proven to reduce fatigue and improve performance, but will require us to revise our notion of the pilot as the heroic, always-on, fully attentive operator to that of an extremely capable human with cognitive and physiological limits.

ALIAS is high-risk and filled with uncertainty, as all DARPA programs are, and it is unclear whether (and how) it might eventually be useful. But as a research program, a kind of advanced thought experiment, it makes explicit what we have been learning for decades: far from a linear progression from human to remote to autonomous aircraft, what we are seeing is a convergence. Human, remote, and autonomous are evolving together, blurring their boundaries.

CHAPTER 7

Autonomy in the Human World

IN THE SUMMER OF 2013, THE MAIN INDUSTRY ASSOCIATION for robotic vehicles, the Association for Unmanned Vehicle Systems International (AUVSI), held its annual convention and trade show in Washington, DC. Here danced robots that trot like pack mules, fly like dragonflies, shoot missiles, and survey crops. But strangely, the big news coming out of the meeting was not new sensors, vehicles, or civilian applications—though these all were present. The big news was a request by the president of the association to the press and public to change the terms of the debate.

"I don't use the word 'drone,'" said AUVSI president Michael Toscano. "There's a Hollywood expectation of what a drone is. Most of it is military; most of it is very fearful, hostile." He was responding to an immediate political threat: outside the convention hall protesters objected to U.S. "drone" strikes and to the prospect of domestic drones snooping on private lives.

"The key word is the word 'system.' That's the word we hope the public will understand," Toscano said. "There is a human being in the system. The human being is what makes the system. When you say the word 'drone,' you don't think of a human being in control."

Toscano was clearly trying to buff up the public image of his industry, which is concerned that public fears of surveillance are holding back the deployment of technologies in U.S. airspace. But this unusual request by an industry to stop calling their products one thing and start calling them something else just drew attention to the importance— and the confusion—surrounding words in the field of mobile robotics.

"Drones" and their work are human products, not impersonal technologies. Unmanned vehicles don't have people on board, but they still embody human efforts. Autonomous vehicles regularly connect with, and return to, the human world. Human operators are linked to vast networks of data, colleagues, and imagery even as they become enmeshed in the details of human events that unfold halfway across the globe.

A basic false premise is that direct, human presence provides authentic, expert experience while robots are unmanned and do the work themselves. We have not taken on debates from artificial intelligence about whether machines might ever think on their own. It is true that an autonomous system might use software that is nondeterministic (i.e., unpredictable), or might employ emergent properties driven by its environment or engage in learning behaviors. Yet any supposedly intelligent system was programmed by people and embeds their world views into the machine.

For the twenty-first century, then, autonomy is *human action removed in time*. This, in a sense, is the essence of the term "programming"— telling the computer what to do at some point in the future, when the program is run. Of course the machine will respond to its environment, and may encounter novel situations, and may even develop unexpected behaviors. But the constraints on those behaviors are still very tight, and very much pre-scripted by the designers and programmers. A rover on Mars may "learn" about its environment, and not try to spin its wheels in the same soil twice if it got stuck the first time, but it would be unable (for both mechanical and software reasons) to "learn" to open a jewelry box if it suddenly came across one.

Still, the fully autonomous robot making its way through a land-scape under computer control remains an attractive idea for engineers. Perceiving the environment, classifying, matching it up to models or prior experience, and making plans to move forward resemble our daily acts of living. Uncertainties in the world (and within the machines), the unexpected that will always foil prior assumptions, make the problem not only harder, but even more interesting. Thinking these problems through, aided by the medium of technology, is a noble effort, engineering at its philosophical best: How do we observe, decide, and act in the world? How do we live with uncertainty?

But we should not confuse technical thought experiments with what's useful in a human context. When lives and resources are at stake, time and time again, over decades, from the deep ocean to the outer planets, we have reined in the autonomy. It is not a story about progress—that one day we'll get it right—but a story about the move from laboratory to field. That transition tempers autonomy, whether the task is to respond to instructions and return useful data or to protect or defend human life.

In retrospect, Neil Armstrong's last-minute manual intervention—turning off the automation of his moon landing—signaled the limits of the twentieth-century vision of full autonomy, and foretold the slow

advent of potent human collaboration and presence. The lone, autonomous drone is as much an anachronism as is the lone, unconnected
computer. The challenges of robotics in the twenty-first century are
those of situating machines within human and social systems. They are
challenges of relationship.

Author Matthew Crawford abandoned his career at a think tank and
set up shop repairing antique motorcycles. He saw it as a matter of
"skilled and active engagement" with the work versus manipulating
abstractions far removed from their anchoring in the material world.
For him, the life of the craftsman appealed because of its accountability
both to the obdurate physical world of the machines and to the social
relationships of customers, mentors, and like-minded motorcycle riders.
For him, a common devotion to the excellence of the work at hand unites
everyone in a working community. "The narrow mechanical things I
concern myself with," writes Crawford, "are inscribed within a larger
circle of meaning; they are in the service of an activity that *we* recognize
as part of a life well lived."

Crawford's career choice seems to speak for the oceanographers who
resisted remote presence, the astronauts who find in physical human
presence the essence of exploration, even the fighter pilots who look
down on Predator pilots as Nintendo warriors. Indeed, technology is
often deployed in the workplace to push labor to be "scaled up, deperson-
alized, and made to answer to forces remote from the scene of work." By
contrast, in Crawford's view, the trades "resist this tendency for remote
control, because they are inherently situated in a particular context."

Ironically, Crawford's conclusion engages exactly the issue we have
been sorting through here: autonomy. He opposes the disconnected
individual, a "sovereign self, unencumbered by attachments to others

and radically free." His work as a motorcycle craftsman connects him to others. His critique of autonomy, while directed at people, resonates with our critique of utopian autonomy in robotics:

> The idea of autonomy denies that we are born into a world that existed prior to us. It posits an essential aloneness; an autonomous being is free in the sense that a being severed from all others is free. To regard oneself this way is to betray the natural debts we owe the world, and commit the moral error of ingratitude. For in fact, we are basically dependent beings: one upon the other, and each on a world that is not of our own making.

Our studies of extreme environments have similarly shown that full autonomy is the less ambitious and less useful problem. More challenging, and more worthwhile, is the problem of autonomy in human settings. How can we design automation that aids operators and supports their skills and identities? How can robots employ autonomous behaviors while still appearing simple, predictable, and transparent to their human collaborators? Will people come to trust unmanned systems when lives are on the line? How should robots be situated within human relationships of power, language, and identity? Even in extreme environments we see the essentially connected nature of machines.

None of this is to say autonomy will not improve, that better sensors, computers, algorithms, and mechanisms won't help us make sense of our world, and help machines live more reliability within it. Quite the contrary, robots still hold great promise as extenders and expanders of human experience. But we will see *through* the rovers, explore *inside* the data, experience presence *in* remote environments, rather than abdicate to machines. New oceanographers, new Martian geologists, even new air force pilots experience fresh worlds through the medium of their machines.

We can find signs of such approaches throughout the research landscape. As I mentioned in previous chapters, DARPA has ambitious programs afoot to engineer human/machine teams, using the full potentials of current computers and robotics. Similarly, the Office of Naval Research, one of the oldest and best-respected federal sponsors of research, has a program in the "science of autonomy." The program includes rigorous mathematical modeling disciplines, like optimization and game theory, control theory, and graph theory. But it also includes human-in-the-loop controls, economics, cognitive psychology, and human-factors engineering. One day soon such sponsors may explore in even greater depth the social dimensions of human behavior, cultural relationships, and knowledge creation.

We have told stories of three kinds of machines: *human operated*, like the airliners we fly every day, *remotely operated*, like the ROVs that explored the *Titanic* and mapped ancient shipwrecks, and *autonomous*, like the undersea vehicle ABE or the Mars rovers, which connect to people less directly and make some decisions on their own. We have followed the people and the machines through larger systems and networks. In each case, human decisions, presence, and expertise are still there but shifting with new technologies, although not always in the ways we expect. It is not the robots themselves, but the novel mixtures of human and automated machines that are changing the nature of the work and the people who do it.

Some still see a linear progression here, as though the most "advanced" technology is the most fully automated or autonomous, with the least human input. Instead our explorations have shown how the three modes evolve together, feeding back and cross-pollinating. The new *Alvin* includes software developed for autonomous vehicles; airliners resemble telerobots that you sit in. The Apollo lunar landings were not simply flown by heroic pilots, but tightly linked to ground controllers and software algorithms that controlled significant parts of the mission. The International Space Station, too, though it houses astronauts, is today largely operated remotely by engineers on the ground.

Military pilots of Predator and Reaper drones fire missiles and kill people from thousands of miles away. But how different is their work from that in the highly computerized cockpits of modern fighter planes, whose pilots rarely see their enemies anywhere except on radar screens? Stark dichotomies between human and not-human, manual and automated are vestiges of the twentieth century.

New approaches are blurring the lines. Technologies like heads-up displays and synthetic vision make flying airplanes more computerized, but less fully automated. Pilots are coupled to the machines in ways that seek to enhance safety. Whether they achieve that promise, of course, depends on how effectively the machines (and their programmers) can account for human roles, and how pilots, and the systems they work in, adapt. It takes more sophisticated technology to keep the humans in the loop than it does to automate them out.

Extreme environments are but harbingers of more earthly domains. As automobiles become more automated, they will reshape the most common technical task that ordinary people perform, challenging us to redefine what it means to be a driver. In medicine, robotically assisted surgery is already changing what it means to be a surgeon—and to be a patient. Factory workers, architects, writers, even race car drivers are all seeing aspects of their roles moving from human minds and bodies into machines—and hence into the domains of other people, who build and program them. The anxiety is real, as are the job losses and social displacements. My own university has mounted a major push to transform education by distributing courses through networks. To what degree, we all ask, does teaching require physical presence? As each of these realms imports robotic and automation technologies, they will also import social changes, complexities, and even accidents.

Our explorations of extreme environments enable us to temper both the naïve promises and the naïve fears currently offered for autonomous machines. Then we can move the conversation (and the creativity) toward questions of human roles, social interaction, and the challenges of reliability and trust. These sit at the core of new conversations on situated autonomy—Where are the people? Who are they? What are they doing? When? Why does it matter?

It is late afternoon on a cold winter day and I am flying my 1993 Beech Bonanza, a single-engine airplane originally designed in 1947, back home after a long trip south. It has recently snowed in New England, and clouds still hover above the airport, their whiteness mingling with the fresh blanket of snow on the landscape down below.

Is this old-fashioned, seat-of-the-pants lone-eagle flying? Hardly. I rely on computerized GPS units that precisely track my location, a digital engine monitor that logs tens of parameters, and satellite links that show weather pictures in vivid detail, all presented to me on five different computer displays (including an iPad). Synthetic vision models my flight through a database in real time. Each system has its own quirks, vulnerabilities, software updates and bugs, and I have to manage them all, accounting for their weaknesses and failures.

Landing an aircraft with computer aids and synthetic vision.

(PHOTO BY JOHN TYLKO)

An older generation of pilots might find me hopelessly digital, artificially divorced from the essence of flight. But I fly in a new world: these devices connect my work to the broader landscape, situating it within networks of great importance for my task and for my safety.

I'm being directed by a voice on air traffic control, navigating via a global satellite network that people are continually maintaining and testing, and landing at an airport maintained by government funds. My airplane flies within these networks connected by multiple, ever-changing communications channels.

Earlier on this flight, one of the cylinders in the engine started indicating an abnormally low temperature. Soon thereafter the computer displaying engine horsepower started reading about 10 percent low, a real cause for concern. The engine sounded fine, however, and the airplane's speed was normal. After some initial concern, and thinking things over, I concluded the engine temperature sensor had gone bad, and there was no real problem in the engine, so the flight could continue. This is a perfectly ordinary, even expected failure.

Perhaps in a few years I will do the same flight with a robot in the seat next to me, and a few years after that the robot may be built into the plane. Would an autonomous algorithm have concluded the engine was about to fail and commenced an emergency landing when none was necessary? Then again, I sure wouldn't have minded talking to someone flying with me from a remote workstation to help sort through the conflicting data.

To approach and land through the clouds, I will fly down the ILS, the radio beam extending from the runway. Usually, I set up the autopilot to fly the approach, lock it onto the beam, and monitor it down to minimums until I take over and manually land the last few hundred feet.

On this approach, about twenty miles from the airport I turn off the autopilot and fly by hand. The autopilot will fail someday, and I need to stay proficient at the skill (if I do not fly every week I feel noticeably rusty). But also, I just enjoy the feeling of being directly in control of the aircraft, creating the smooth, precise motions that get me home.

Over the radio, the disembodied voice of air traffic control gives me some directions to fly to line up for the beam. When I'm almost there, he clears me for the approach, meaning I am freed to fly under my own direction. On my computer display, the graphic "needle" indicating the received radio beam comes alive. I follow the needle onto the final approach course; when its vertical partner, the glide slope, comes alive, I lower the wheels and the airplane begins to descend. I read through a checklist, a procedure I composed months before in the comfort of an evening at home.

With minute hand movements I adjust the airplane's pitch to stay on the glide slope. Though I can't yet see anything out the window, my computer display shows a synthetic view of the terrain down below. A green flight path vector is overlaid on the terrain to show where the airplane is going. By gently maneuvering the airplane, if I put the vector on the end of the virtual runway, the airplane will go right to the landing zone.

A few hundred feet above the ground, I come down from the clouds. Suddenly appearing right in front of me is the runway. I gently ease back on the throttle, and then on the control yoke. With a muffled squeak, the airplane's tires kiss the runway; a little braking slows it down and I am home. It is an immensely satisfying experience. The total focus it requires leaves me feeling tired, but relaxed and refreshed.

ACKNOWLEDGMENTS

As a research project this book was about seven years in the making, but as an intellectual journey it has been more than twenty-five. Numerous friends and colleagues contributed along the way.

Tim Cullen's dissertation research proved central to my arguments and data in Chapter 4. Ray O'Mara provided the perspective of a skeptical fighter pilot turned scholar. Postdocs Zara Mirmalek and Yanni Loukissas helped through many conversations; Zara contributed to the research that went into Chapter 3.

For part of the research I spent a sabbatical year as a visiting scientist at Aurora Flight Sciences, at their Engineering Research Center in Cambridge, Massachusetts. Aurora colleagues not only allowed me to study their projects but also included me in efforts to prove new approaches to problems of humans and autonomy: John Langford, Jim Paduano, John Wissler, Javier de Luis. John Tylko of Aurora and MIT provided an expansive entrée into the business of autonomy.

Ed Hutchins kindly hosted an early presentation of the ideas in this book at UCSD. Nat Sims and I have had numerous conversations, especially about the overlapping relationships between people and machines in aviation and medicine. I'm grateful for the advice of Tom Imrich on flying airliners. As the book took shape, students in my "Human, Remote, and Autonomous Systems" graduate seminar at MIT helped work through the material over two consecutive years, especially: Mark Boyer, Madeline Elish, Steve Fino, Sherrie Hall, Mark Harris, Kyle Kotowick, Scott Nill, Matt Rabe, Jason Ryan, Sathya Silva, Aleksandra Stankovic, and Erik Stayton.

A number of friends were kind enough to read through the manuscript and offer incisive, critical readings: Kathy Abbott, Robert Ballard, Bill Clancey, Frank Levy, John Markoff, Vic McElheny, Bob Moreau, Don Norman, Chuck Oman, Wade Roush, Rosalind Williams, Dana Yoerger. Each of their readings was part of a longer conversation of the type that makes intellectual life worth thinking.

The Jay Last Fund in the Science, Technology, and Society Program at MIT provided essential funding to support research and writing for several summers. The L. Dennis Shapiro Fund provided financial support to the Laboratory for Automation, Robotics, and Society (LARS), which conducted several of the research projects that composed the book. Dennis Shapiro himself has provided much more than financial support, as a pilot, mentor, and friend. The Alfred P. Sloan Foundation program for the Public Understanding of Science and Technology, under Program Director Doron Weber, supported the final stages of writing. Sally Chapman at MIT provided invaluable assistance for the illustrations.

I also thank the many interviewees at the places I have called EuroAir and HudView and at the Woods Hole Oceanographic Institution.

My agent, Katherine Flynn, understood the potential of this book, even when she first saw it as a muddy proposal, and helped improve it immeasurably. She led me to my editor at Viking/Penguin, Melanie Tortoroli, who has always shared my vision for the book and contributed much to the final product.

My daughters, Lucia and Clara, whose lives began and flowered during my work on this book, will hopefully read it someday. The book and my wife Pamela's choral conducting have been parallel family projects. I only hope the sounds in here are a fraction as beautiful.

NOTES

CHAPTER 1: HUMAN, REMOTE, AUTONOMOUS

2 **a team of twelve engineers:** This account is based on the author's interviews with Mike Purcell, Woods Hole Oceanographic Institution, August 2011.

4 **"only one software upgrade away":** "Terminate the Terminators," *Scientific American* 303, no. 1 (July 2010): 30.

5 **In the domain of work:** Frank Levy, *The New Division of Labor: How Computers Are Creating the Next Job Market* (New York: Russell Sage Foundation; Princeton, NJ: Princeton University Press, 2004). Erik Brynjolfsson and Andrew McAfee, *Race Against the Machine: How the Digital Revolution Is Accelerating Innovation, Driving Productivity, and Irreversibly Transforming Employment and the Economy* (Lexington, MA: Digital Frontier Press, 2012). Illah Reza Nourbakhsh, *Robot Futures* (Cambridge, MA: MIT Press, 2013).

8 **"this concept of keeping the human in the loop":** Peter W. Singer, *Wired for War: The Robotics Revolution and Conflict in the Twenty-First Century* (New York: Penguin, 2009).

10 **"there are no fully autonomous systems":** Defense Science Board, "Task Force Report: The Role of Autonomy in DoD Systems," Washington, DC: Office of the Under Secretary of Defense for Acquisition, Technology, and Logistics, July 2012: 33.

12 **One recent report introduces the term "increasing autonomy":** *Autonomy Research for Civil Aviation: Toward a New Era of Flight* (National Research Council, 2014).

CHAPTER 2: SEA

24 **The Skerki D survey:** Robert D. Ballard et al., "The Discovery of Ancient History in the Deep Sea Using Deep Submergence Technology," *Deep-Sea Research I* 47 (2000): 1591–1620.

27 *Alvin* **was part of a broad landscape:** Frank Busby, *Undersea Vehicles Directory 1987* (Arlington, VA: *Busby Associates Inc.,* 1987). Richard Geyer, ed., *Submersibles and Their Use in Oceanography and Ocean Engineering,* Elsevier Oceanography Series 17 (Amsterdam; New York : Elsevier Scientific Publications Co., 1977).

28 **"scientific research salesman":** Robert D. Ballard and Will Hively, *The Eternal Darkness: A Personal History of Deep-Sea Exploration* (Princeton, NJ: Princeton University Press, 2000), 63.

29 **"It was fortunate that Alvin was there":** Frank Taylor interview with Robert Ballard, April 27, 2000, Woods Hole, MA, Woods Hole Oceanographic Institution Archives.

29 **much of the scientific community:** Ballard and Hively, *The Eternal Darkness,* 158

29 **Ballard developed methods:** Frank Taylor interview with ANGUS group, February 27, 2002, Woods Hole, MA, Woods Hole Oceanographic Institution Archives.

29 **"And the nimble little white submarine":** Ballard and Hively, *The Eternal Darkness,* 49.

30 **"We asked Holger and Fred how to proceed":** Ibid., 186.

32 **a computer connection to the ship's control system as well:** Frank Taylor interview with ANGUS group, February 27, 2002. Woods Hole Oceanographic Institution Archives, 13.

32 **"We found running the sled":** Ibid., 14.

33 **In team member Steve Gegg's words:** Ibid., 21.

37 **Ballard split away from the** *Alvin* **group:** Ballard and Hively, *The Eternal Darkness,* 233.

38 **"He was very much interested in social issues":** Author interview with Dana Yoerger, Woods Hole, MA, August 2011.

41 **"an improbable kite of white steel":** Ballard and Hively, *The Eternal Darkness,* 8.

42 **"Our scanning human eyes":** Ibid., 9, 240–41.

45 **"the** *Alvin* **group lost some of its popular glamour":** Ibid., 295

46 **"***Alvin* **became a different machine":** Author interview with Will Sellars, Woods Hole, MA, August 2011.

46 "We went from [testing in] ten feet of water": Frank Taylor interview with Martin Bowen, October 2001, Woods Hole, MA, Woods Hole Oceanographic Institution Archives, 35.

46 the cook on *Atlantis II* made a special cake for Ballard: Ballard and Hively, *The Eternal Darkness*, 297.

46 "bureaucratic inflexibility": Ibid., 301, 312.

47 "When the *Alvin* pilot turned on": Frank Taylor interview with Martin Bowen, October 2001, Woods Hole, MA, Woods Hole Oceanographic Institution Archives, 35.

47 "in the portholes you'd see an eight-story building": Ibid., 40.

48 "all I could see was that big gaping hole": Author interview with Dudley Foster, Woods Hole, MA, August 2011.

49 "I was in that vehicle's eye": Frank Taylor interview with Martin Bowen, October 2001, Woods Hole, MA, Woods Hole Oceanographic Institution Archives, 38–39.

49 "I was just flying this thing": Ibid., 40–42.

49 "As we sat inside *Alvin*": Ibid., 43.

50 Will Sellars was amazed: Author interview with Will Sellars, Woods Hole, MA, August 2011.

50 The *Alvin/Jason Jr.* combination was the feature of a new *National Geographic*: *National Geographic* cover, December 1986.

51 after *Titanic*, it was never used again: Martin Bowen and I rebuilt *Jason Jr.* and redid its internal electronics in 1991, in preparation for an expedition to the Galapagos. Unfortunately, the barge carrying *JJ*, and all of our equipment, sank while being towed to the islands. *JJ* sits today in a crate, inside a shipping container, three miles down in the South Pacific. Ironically, the only piece of equipment likely to be undamaged is *JJ* and its titanium pressure housing.

55 "It's a ballet": Frank Taylor interview with Martin Bowen, October 2001, Woods Hole, MA, Woods Hole Oceanographic Institution Archives, 45.

56 "You become overwhelmed with input": Author interview with Will Sellars, Woods Hole, MA, August 2011.

56 "With robots you could have a whole gallery of experts": Frank Taylor interview with Martin Bowen, October 2001, Woods Hole, MA, Woods Hole Oceanographic Institution Archives, 44.

57 "I just starting mapping things in my own head": Ibid., 67.

60 "People would say, ROVs?": Ibid.

60 Another member of the team recalled: Interviewee unattributed by request.

61 "And you find people that go": Author interview with Will Sellars, Woods Hole, MA, August 2011.

CHAPTER 3: AIR

69 **"total loss of cognitive control of the situation":** *Final Report on the Accident on 1st June 2009 to the Airbus A330-203 Registered F-GZCP Operated by Air France Flight AF 447 Rio de Janeiro–Paris.* Bureau d'Enquetes et d'Analyses, July 2012: 217.

71 **de Crespigny was forced to go deeper:** Richard de Crespigny, *QF32* (Macmillan Australia, 2012). Australian Transport Safety Bureau, "In-flight Unconfined Engine Failure Overhead Batam Island, Indonesia, 4 November 2010, VH-OQA, Airbus A380–842," *ATSB Transport Safety Report, Aviation Safety Occurrence Investigation—AO-2010-089 Final 27 June, 2013.*

73 **"syntax, sequence, and procedure":** Robert Moreau, personal communication with author, December 2014.

73 **what happened on Air France 447:** For a journalistic summary of the accident, see William Langewiesche, "The Human Factor," *Vanity Fair,* October 2014.

73 **summarized the results this way:** J. K. Lauber quoted in Nadine Sarter et al., *Cognitive Engineering in the Aviation Domain,* 1st edition (CRC, 2000), 275–76.

75 **a joint industry-FAA working group:** PARC/CAST Flight Deck Automation Working Group, "Operational Use of Flight Path Management Systems," Federal Aviation Administration, September 5, 2013.

76 **"We're not eliminating human error":** Kathy Abbott, personal communication with the author, November 2013.

77 **"The twentieth century was born yearning for a new type of hero":** Robert Wohl, *A Passion for Wings: Aviation and the Western Imagination, 1908–1918* (New Haven: Yale University Press, 1994), 29.

78 **The story of the pilot in the twentieth century:** Wohl, *A Passion for Wings,* 30.

78 **"If you can't fly without looking at your airspeed":** Beryl Markham, *West with the Night* (New York: North Point Press, 2013).

79 **Doolittle flew the first instrument flight in 1929** James Harold Doolittle and Carroll V. Glines, *I Could Never Be So Lucky Again* (New York: Bantam Books, 1992). Richard Hallion, *Legacy of Flight: The Guggenheim Contribution to American Aviation* (Seattle: University of Washington Press, 1977). Erik M. Conway, *Blind Landings: Low-Visibility Operations in American Aviation, 1918–1958* (Baltimore: Johns Hopkins University Press, 2006).

79 **intentionally called them "instruments":** Michael Aaron Dennis, "A Change of State: The Political Cultures of Technical Practice at the MIT Instrumentation Laboratory and the Johns Hopkins University Applied Physics Laboratory, 1930–1945," PhD dissertation, 1990. Donald A. MacKenzie, *Inventing Accuracy: A Historical Sociology of Nuclear Missile Guidance* (Cambridge, MA: MIT Press, 1993).

80 **Buck's memoir of the ensuing forty years:** Robert Buck, *North Star over My Shoulder: A Flying Life* (New York: Simon & Schuster, 2002). Robert Buck, *The Pilot's Burden: Flight Safety and the Roots of Pilot Error*, 1st edition (Ames: Iowa State University, 1994).

81 **began to replace the terms "pilot" and "copilot":** Robert Daley, *An American Saga: Juan Trippe and His Pan Am Empire* (Riviera Productions Ltd., 1980; Kindle edition, 2010) location 2332.

81 **airlines began sending licensed mechanics along:** Nick Komons, *The Third Man: A History of the Airline Crew Complement Controversy, 1947–1981* (Washington, DC: Department of Transportation, Federal Aviation Administration, 1987), 12.

81 **U.S. government began requiring professional flight engineers:** Ibid., 12.

82 **During congressional hearings they presented photographs:** Ibid., 48.

82 **"knobs, dials, and gauges vanished from the cockpit":** Ibid., 37

87 **"He is involved but detached":** Richard Collins, "Look! No Hands!," *Flying* (January 1986): 73–75.

87 **Richard de Crespigny recalled the autoland:** De Crespigny, *QF32*.

88 **"Watching the precise performance":** Collins, "Look! No Hands!"

88 **according to the 2013 FAA working group on automation:** Flight Deck Automation Working Group, "Operational Use of Flight Path Management Systems," Federal Aviation Administration, September 5, 2013.

95 **Thomas is an advocate for HUDs:** Interviews with HUD users were conducted under a human subjects research protocol. Though many of the subjects permitted their real names to be used, in keeping with ethnographic conventions I am using pseudonyms for pilots, engineers, and company names.

98 **One study found that the pilot's stresses when using autoland:** Daniel Bandow, "Head Up Guidance System Model 2100 and Human-Machine Interaction," DBS Systems.

99 **"Since the HUD is flight path-centric":** Robert Moreau, personal communication with author, December 2014.

105 **In 2009 the Flight Safety Foundation:** Flight Safety Foundation, "Head-up Guidance System Technology: A Clear Path to Increasing Flight Safety," Special Report, November 2009.

106 **The Asiana crew had no heads-up display:** National Transportation Safety Board Investigative Hearing: Crash of Asiana Flight 214 San Francisco 7/6/2014, archived at: http://www.youtube.com/watch? v= 9X-gmagrMjs.

106 **its visual equivalent, a system of fixed runway lights:** These lights use lenses to project red and white light beams at different angles into the sky. If the pilot

is too high, the beams will appear white, too low and they appear red. Red and white means the aircraft is on the glide slope.

106 **The Asiana pilot flying said he was stressed about landing:** National Transportation Safety Board, *Descent Below Visual Glidepath and Impact With Seawall, Asiana Airlines Flight 214,* San Francisco, California. July 6, 2013. NTSB Number: AAR1401, June 24, 2014: 99.

107 **"Without greater opportunity for pilots to manually fly the airplane":** NTSB Asiana flight 214 report, 62–63, 102–3. In the other 5 percent of landings the type was not recorded.

107 **Some of these innovations have been called "information automation":** Charles E. Billings, *Aviation Automation: The Search for a Human-Centered Approach,* Human Factors in Transportation series (Mahwah, NJ: Lawrence Erlbaum Associates, 1997).

CHAPTER 4: WAR

113 **transported into a remote war zone:** This account, and all ethnographic descriptions and quotes in this chapter, are drawn from Timothy M. Cullen, *"The MQ-9 Reaper Remotely Piloted Aircraft: Humans and Machines in Action,"* Massachusetts Institute of Technology dissertation, 2011. Cullen's dissertation, after it had been accepted by a faculty committee (which I chaired) at MIT, though it contained no classified information, was heavily redacted by the air force before being placed in the MIT archives. The MIT committee accepted the action on the condition it could be summarized in this book.

117 **public presentations and memoirs:** Matt Martin and Charles W. Sasser, *Predator: the Remote-Control Air War over Iraq and Afghanistan: A Pilot's Story* (Minneapolis, MN: Zenith Press, 2010). Martin was an active duty air force officer when he cowrote this memoir, which means he had to have it cleared by the service before publication. Therefore this account can be considered a quasi-official account of a Predator operator, providing some insight into what the air force would like to present as the Predator operator's experience. Nonetheless, because of Martin's public persona, at least some in the Predator community consider him a pariah or, in the words of one officer, "a leper."

118 **"We shouldn't have pilots stick-and-ruddering UAVs":** Houston R. Cantwell, "Operators of Air Force Unmanned Aircraft Systems: Breaking Paradigms," *Air & Space Power Journal* (Summer 2009): 70.

120 **the air force put more RPA operators:** United States Air Force, RPA Vector: *Vision and Enabling Concepts 2013–2038,* 2014: 18.

120 **a CAP requires more than a hundred and fifty people:** J. R. Gear, "USAF RPA Update: Looking to the Future," U.S. Air Force Briefing Slides, June 3, 2011. Unclassified.

120 **the symbol of all that is wrong with American technological power:** Bradley Jay Strawser and Jeff McMahan, *Killing by Remote Control: The Ethics of an Unmanned Military* (Oxford; New York: Oxford University Press, 2013). Ronald Arkin, *Governing Lethal Behavior in Autonomous Robots*, 1st edition (Boca Raton, FL: Chapman and Hall/CRC, 2009). For some of the more thoughtful voices in the debate, see "The Three Faces of Drone War," TomDispatch.com.

120 **Predator has been a focal point:** In what follows I will use "Predator" to refer to both aircraft, as "Reaper" is technically "Predator B," though I will discuss some of the differences.

121 **"The Nintendo mentality is a detached mentality":** Chris Cole, Mary Dobbing, and Amy Hailwood, "Convenient Killing: Armed Drones and the 'Playstation' Mentality," The Fellowship of Reconciliation, Oxford, UK: September, 2010.

121 **"peripheral systems locked in a seemingly endless, inglorious loop":** Thomas P. Ehrhard, *Unmanned Aerial Vehicles in the United States Armed Services: A Comparative Study of Weapon System Innovation* (Baltimore: Johns Hopkins University Press, 2000), 16.

122 **their arrow-straight trajectories:** David Mindell, *Between Human and Machine: Feedback, Control, and Computing Before Cybernetics* (Baltimore: Johns Hopkins University Press, 2000).

124 **A few even flew in Iraq in 2003:** "Ryan Firebee," Wikipedia, https://en.wikipedia .org/wiki/Ryan_Firebee, accessed June 16, 2015.

124 **little evidence of this "white scarf syndrome":** Ehrhard, *Unmanned Aerial Vehicles,* 41.

125 **"only the obscure novelty of a mechanical feat":** Ibid., 652, 674.

127 **Karem said in a recent interview:** Mark Harris, "The Dronefather," *The Economist* 405, no. 8813 (December 2012). Also see Richard Whittle, "The Man Who Invented the Predator," *Air & Space* (April 2013).

127 **Jumper saw real-time video feeds:** Thomas P. Ehrhard and General Billy Mitchell Institute for Airpower Studies, *Air Force UAVs: The Secret History* (Arlington, VA: Mitchell Institute Press, 2010), 49–50. Jon Jason Rosenwasser and Fletcher School of Law and Diplomacy, *Governance Structure and Weapon Innovation: The Case of Unmanned Aerial Vehicles* (Medford, MA: Tufts University, 2004), 256.

128 **benefitted from the new ubiquity of GPS:** Ehrhard, *Unmanned Aerial Vehicles,* 41, 185.

128 **The crews began to refer to the phenomenon as "Predator porn":** Richard Whittle, *Predator: The Secret Origins of the Drone Revolution*, 1st edition (New York: Henry Holt and Co., 2014), 115, 128.

129 **this secretive outfit:** Bill Grimes, *The History of Big Safari* (Bloomington, IN: Archway Publishing, 2014).

132 **Training and employment standards:** Cullen, "The MQ-9 Reaper," 216.

133 **The pilots monitored their timing and flight paths:** William B. O'Connor, *Stealth Fighter: A Year in the Life of an F-117 Pilot* (MBI Publishing Company LLC, 2012). O'Connor does not draw attention to the fact that the aircraft was under computer control during the bombing run, but it is implicit in the narrative.

133 **Nonrated pilots still had to have civilian commercial pilot ratings:** Wayne Chappelle, Kent McDonald, and Katharine McMillan, "Important and Critical Psychological Attributes of USAF MQ-1 Predator and MQ-9 Reaper Pilots According to Subject Matter Experts." Air Force Research Laboratory 711th Human Performance Wing, May 2011. AFRL-SA-WP-TR-2011-0002. (USAF), U.S. Air Force, USAF Accident Investigation Board, World Spaceflight News, U.S. Military, and Department of Defense (DoD), *U.S. Air Force Aerospace Mishap Reports: Accident Investigation Boards for UAV/UAS Remotely Piloted Aircraft (RPA) Incidents Involving the MQ-1B Predator in Afghanistan, Iraq, and California* (Progressive Management Publications, Kindle edition, 2012), location 1745–61.

135 **One early Predator pilot was shocked to find:** Whittle, *Predator*, 96–100.

135 **a "dialogue of the deaf":** Houston R. Cantwell, *Beyond Butterflies: Predator and the Evolution of Unmanned Aerial Vehicles in Air Force Culture* (n.p.: Biblioscholar, 2012), 25.

137 **exchange video and voice communications directly with CIA:** National Commission on Terrorist Attacks, *The 9/11 Commission Report: Final Report of the National Commission on Terrorist Attacks Upon the United States* (New York: W. W. Norton & Company, 2004), 189–90. Whittle, *Predator*, 151–61.

138 **On their seventh flight, on September 27:** Whittle, *Predator, 151–61*.

138 **He ordered Big Safari to begin arming Predator:** Grimes, *The History of Big Safari*, 332.

138 **was seen as politically and legally troubling:** Whittle, *Predator*, 211, 222. *9/11 Commission Report*, 211–12.

139 **Confusion between the CIA and the air force:** Whittle, *Predator*, 243–53.

142 **One 2011 study:** Joseph Ouma, Wayne L. Chappelle, and Amber Salinas, *Facets of Occupational Burnout Among U.S. Air Force Active Duty and National Guard/*

Reserve MQ-1 Predator and MQ-9 Reaper Operators. Air Force Research Laboratory Report AFRL-SA-WP-TR-2011-0003, 2011: 11–12.

142 **in the 1970s the "guy in back" was eliminated:** Steven A. Fino, "Flying Knights
or Flying Scientists? A Cognitive History of the US Air Force Fighter Pilot in
Air-to-Air Combat, 1950–1980," PhD Dissertation, Massachusetts Institute of
Technology, 2014.

143 **"For every disgruntled [Predator] pilot hanging on:** Linda Shiner, "Predator,"
Air & Space Magazine (April–May 2001): 48, www.airspacemag.com/military
-aviation/predator-first-watch-2096836/, accessed February 4, 2014.

143 **Crews then coordinate via text messages:** Cantwell, *Beyond Butterflies,* 28.

145 **Predator crews at first added a FalconView display:** Cullen, "The MQ-9 Reaper,"
258. Jon R. Lindsay, "'War upon the Map': User Innovation in American Military Software," *Technology and Culture* 51, no. 3 (2010): 619–51.

146 **Predator crews would routinely monitor . . . separate conversations:** Cullen,
"The MQ-9 Reaper," 257.

146 **In the words of David Deptula:** David Deptula, "Drones Best Weapons We've
Got for Accuracy, Control, Oversight," Breaking Defense Web site, http://
breakingdefense.com/2013/02/retired-gen-deputula-drones-best-weapons
-weve-got-for-accurac, accessed May 19, 2014.

147 **users reported seeing messages like "concur" or "shoot now":** Cullen, "The
MQ-9 Reaper," 258.

148 **Predator crews began to feel like "chat-activated sensors":** Cullen, "The MQ-9
Reaper," 259.

150 **the presence is through an American lens:** This idea is elaborated in Derek
Gregory, "From a View to a Kill: Drones and Late Modern War," *Theory, Culture
& Society* 28, no. 7–8 (December 1, 2011): 188–215.

150 **has accounted for civilian casualties:** David S. Cloud, "Anatomy of an Afghan
War Tragedy," *Los Angeles Times,* April 10, 2011, http://articles.latimes.com/2011/
apr/10/world/la-fg-afghanistan-drone-20110410, accessed January 9, 2015.

151 **the air force prohibited Predator crews from using the term:** Quoted in Cloud,
"Anatomy of an Afghan War Tragedy."

151 **those who have looked at civilian casualties:** Avery Plaw, "Counting the Dead:
The Proportionality of Predation in Pakistan," in Strawser and McMahan,
Killing by Remote Control, Chapter 7.

152 **A study of stress and burnout among Predator operators:** Ouma et al., *Facets
of Occupational Burnout Among U.S. Air Force Active Duty and National Guard/
Reserve MQ-1 Predator and MQ-9 Reaper Operators,* 1.

153 **reported to GQ magazine:** Heather Linebaugh, "I Worked on the US Drone Program: The Public Should Know What Really Goes On," *The Guardian*, December 29, 2013, http://www.theguardian.com/commentisfree/2013/dec/29/drones-us-military. Matthew Power, "Confessions of a Drone Warrior," *GQ* magazine, March 2013. Elisabeth Bumiller, "Drone Pilots, Waiting for a Kill Shot 7,000 Miles Away," *New York Times*, July 29, 2012, http://www.nytimes.com/2012/07/30/us/drone-pilots-waiting-for-a-kill-shot-7000-miles-away.html.

153 **A 2013 air force study found:** Jean L. Otto and Bryant J. Webber, "Mental Health Diagnoses and Counseling among Pilots of Remotely Piloted Aircraft in the United States Air Force," *MSMR* 20, no. 3 (March 2013): 3–8.

153 **The air force responded:** Lee Ferran, "Drone 'Stigma' Means 'Less Skilled Pilots' at Controls of Deadly Robots," ABC News, April 29, 2014, http://abcnews.go.com/Blotter/drone-stigma-means-skilled-pilots-controls-deadly-robots/story?id=23475968 accessed May 19, 2014.

154 **Many pilots point out the benign air environment:** Dan Hampton, *Viper Pilot: A Memoir of Air Combat* (New York: HarperCollins, 2012).

155 **Blair had served as a pilot of both a C-130 gunship and a Predator:** Dave Blair, "Ten Thousand Feet and Ten Thousand Miles: Reconciling Our Air Force Culture to Remotely Piloted Aircraft and the New Nature of Air Combat," *Air & Space Power Journal* 26, no. 3 (May–June 2012): 61–69. On this issue also see Robert Sparrow, "War Without Virtue," in Strawser and McMahan, *Killing by Remote Control*, Chapter 5.

156 **"No way is a UAV pilot [sitting] in a box":** Blair, "Ten Thousand Feet."

156 **forced the air force to examine its notions of "airmanship":** Cantwell, "Operators of Air Force Unmanned Aircraft Systems."

156 **In 2009 the air force created a new career classification:** Aaron Church, "RPA Ramp Up," *Air Force* 94, no. 6 (2011): 58–60.

156 **Leon Panetta announced a new decoration:** Leon Panetta, "Distinguished Warfare Medal," Memorandum, February 13, 2013, http://www.defense.gov/news/distinguishedwarfaremedalmemo.pdf, accessed July 23, 2014.

157 **Within weeks, the new medal was canceled:** "VFW Believes Distinguished Warfare Medal Should Not Outrank the Bronze Star, Purple Heart," *VFW—Veterans of Foreign Wars*, http://www.vfw.org/News-and-Events/Articles/2013-Articles/VFW-WANTS-NEW-MEDAL-RANKING-LOWERED/, accessed July 23, 2014. "Military Order of the Purple Heart Opposes Precedence of New Defense Medal," Military Order of the Purple Heart, February 15, 2013, http://www.purpleheart.org/News.aspx?Identity=238, accessed July 23, 2014. "US

Military Announces New Medal for Cyberwarfare and Drone Operation," *The Verge,* http://www.theverge.com/2013/2/13/3985802/us-military-announces -distinguished-warfare-medal-for-cyberwarfare-drones, accessed May 17, 2014. "US Defense Secretary Downgrades Drone Medal after Outcry," *The Verge,* http://www.theverge.com/2013/4/15/4228112/defense-secretary-down grades-drone-medal-distinguishing-device, accessed May 17, 2014. "Distinguished Warfare Medal," *Wikipedia,* May 4, 2014, http://en.wikipedia.org/w/index .php?title=Distinguished_Warfare_Medal&oldid=607013488, accessed May 16, 2014. "Medals for Drone Warriors Canceled," *New York Times,* April 15, 2013, http://www.nytimes.com/2013/04/16/us/politics/medals-for-drone-warriors -canceled.html, accessed July 23, 2014.

157 **"whether there isn't danger enough to give us glory":** William Keeler, quoted in David Mindell, *Iron Coffin: War, Technology, and Experience Aboard the USS Monitor,* 2nd edition (Baltimore: Johns Hopkins University Press, 2012).

CHAPTER 5: SPACE

161 **Most of them said they turned off the digital aids** David Mindell, *Digital Apollo: Human and Machine in Spaceflight* (Cambridge, MA: MIT Press, 2008).

161 **The closest the autoland came to landing the shuttle:** L. B. McWhorter et al., "Space Shuttle Entry Digital Autopilot," SP-2010-3408, NASA Johnson Space- flight Center, 2010. G. Tsikalas, "Space Shuttle Autoland Design," American Institute of Aeronautics and Astronautics, 1982. T. T. Myers et al., "Space Shuttle Flying Qualities and Flight Control System Assessment Study—Phase II," NASA Contractor Report 170406, December 1983. H. Law and L. B. McWhorter, "Shuttle Autoland Status Summary" in *Space Programs and Tech- nologies Conference,* American Institute of Aeronautics and Astronautics, 1992–1273. For comments on later autoland tests that were canceled, see "Breaking Through," Wayne Hale's Blog, http://waynehale.wordpress.com/ 2011/03/11/breaking-through/, accessed July 4, 2014.

163 **observers asked whether people still need to venture into space:** MIT Space, Policy, and Society Working Group, "The Future of Human Spaceflight," white paper, December 2008.

164 **Von Braun included an orbiting space telescope:** Wernher von Braun, "Cross- ing the Last Frontier," *Collier's,* March 22, 1952, 24–25, discussed in H. McCurdy, "Observations on the Robotic versus Human Issue in Spaceflight," in Steven J. Dick and Roger D. Launius, eds., *Critical Issues in the History of Spaceflight,* NASA History Series (Washington, DC: National Aeronautics and Space Administration, 2005): 77–106.

165 **"All the astronaut has to do":** Comments from Cepollina, Hoffman, and Musgrave and mission narrative from "Rescuing Hubble: MIT Aero/Astro Gardner Lecture/Sympoisum," November 13, 2013. Video available online at http://teach ingexcellence.mit.edu/category/must-see/rescuing-hubble, accessed July 3, 2014. Additional comments from Jeffery Hoffman, and oral history interviews with Jennifer Ross-Nazzal, April 2, November 3, November 12, and November 17, 2009. NASA Johnson Space Center Oral History Project.

172 **space observers saw the Hubble as the first "victim":** Daniel Morgan, "Hubble Space Telescope: Should NASA Proceed with a Servicing Mission?" *Congressional Research Service Reports*, January 1, 2006: 3.

173 **His wariness was bolstered by a study:** *Assessment of Options for Extending the Life of the Hubble Space Telescope: Final Report*, http://www.nap.edu/catalog .php?record_id=11169, accessed July 4, 2014.

173 **Arthur Whipple, systems engineer:** Arthur Whipple, "A Comparison of Human and Robotic Servicing of the Hubble Space Telescope." Presentation to Future In-Space Operations (FISO) Teleconference, October 14, 2009.

174 **It used more than a hundred brand-new tools:** Jill McGuire, "HST Crew Aids and Tools: Working in Space Today and Tomorrow." Presentation at the Goddard Space Flight Center, September 14, 2009. For imagery of the mission see Dennis R. Jenkins and Jorge R. Frank, *Servicing the Hubble Space Telescope: Shuttle Atlantis, 2009* (North Branch, MN: Specialty Press, 2009).

175 **"It has unfolded in excruciatingly slow motion":** Steven Squyres, quoted in William J. Clancey, *Working on Mars: Voyages of Scientific Discovery with the Mars Exploration Rovers* (Cambridge, MA: MIT Press, 2012), 129.

176 **"best done by one or two geologists":** Comments by Kip Hodges at Exploration Telerobotics Symposium, NASA Goddard Space Flight Center, May 2–3, 2012, http://telerobotics.gsfc.nasa.gov, accessed July 3, 2014.

176 **Historian Naomi Oreskes points out:** Naomi Oreskes, *The Rejection of Continental Drift: Theory and Method in American Earth Science* (New York: Oxford University Press, 1999).

177 **"A well-trained astronaut can talk just as well as a trained geologist":** Interview with geologist, March 2005, notes in the author's possession.

178 **describes the Apollo work as "really telerobotics":** Comments by Kip Hodges at Exploration Telerobotics Symposium.

178 **Head emphasizes it was also important to turn loose the astronauts:** Jim Head and Dave Scott, discussion with the author at "Engineering Apollo" class at MIT, April 2013.

179 **has recently been working on such vehicles:** Akil J. Middleton, "Modeling and Vehicle Performance Analysis of Earth and Lunar Hoppers," Thesis, Massachusetts Institute of Technology, 2010. P. Cunio et al., "Further Development and Flight Testing of a Prototype Lunar and Planetary Surface Exploration Hopper: Update on the TALARIS Project" in *AIAA SPACE 2010 Conference & Exposition*, American Institute of Aeronautics and Astronautics.

181 **a record for off-earth planetary driving:** "NASA's Long-Lived Mars Opportunity Rover Sets Off-World Driving Record," NASA news release, July 28, 2014.

181 **"planetary jet lag":** Zara Mirmalek, "Solar Discrepancies: Mars Exploration and the Curious Problem of Inter-Planetary Time," PhD Dissertation, University of California–San Diego, 2008.

183 **he became interested in the scientists' experience of presence:** All Clancey quotes from William J. Clancey, "Becoming a Rover," in *Simulation and Its Discontents*, Sherry Turkle, ed. (Cambridge, MA: MIT Press, 2009), 107–27, or William J. Clancey, *Working on Mars: Voyages of Scientific Discovery with the Mars Exploration Rovers* (Cambridge, MA: MIT Press, 2012).

183 **"My body is always the rover":** Quoted in Clancey, "Becoming a Rover," 7, 45, 118. For scientists' bodily involvement in the rovers, see Janet Vertesi, "Seeing Like a Rover: Visualization, Embodiment, and Interaction on the Mars Exploration Rover Mission," *Social Studies of Science* 42 (2012): 393–414.

184 **he often refers to the team itself being on Mars:** Steven Squyres, *Roving Mars: Spirit, Opportunity, and the Exploration of the Red Planet* (New York: Hyperion; London: Turnaround, 2006).

184 **"the slope immediately in front of us":** Squyres, quoted in Clancey, *Working on Mars*, 100. Also see Squyres, *Roving Mars*, 100.

184 **"enabling a feeling of synergistic operation":** Clancey, *Working on Mars*, 58.

185 **a "fundamental fallacy":** Comments by Jim Bell and Jake Bleacher at Exploration Telerobotics Symposium, NASA Goddard Space Flight Center, May 2–3, 2012, http://telerobotics.gsfc.nasa.gov/, accessed July 3, 2014. Also see Clancey, *Working on Mars*, 129–37.

186 **One of MER's robotics engineers was "surprised":** Clancey, *Working on Mars*, 117–21.

188 **"When Congress starts using the phrase 'human presence'":** Dan Lester, "Achieving Human Presence in Space Exploration," *Presence* 22, no. 4 (Fall 2013): 345–49.

188 **Lester and his NASA colleague Harley Thronson argue:** Dan Lester and Harley Thronson, "Human Space Exploration and Human Spaceflight: Latency and the Cognitive Scale of the Universe," *Space Policy* 27, no. 2 (May 2011): 89–93.

CHAPTER 6: BEYOND UTOPIAN AUTONOMY

193 **"I don't have maps this good of Iceland":** Dana Yoerger, interview with the author, Woods Hole, MA, August 2011. D. R. Yoerger, A. M. Bradley, M. H. Cormier, W. B. F. Ryan, and B. B. Walden, "High Resolution Mapping of a Fast Spreading Mid-Ocean Ridge with the Autonomous Benthic Explorer," *11th International Symposium on Unmanned Untethered Submersible Technology (UUST99)*, Durham, New Hampshire, August 1999.

193 **Yoerger and his team developed methods:** Christopher German, Dana R. Yoerger, Michael Jakuba, Timothy M. Shank, Charles H. Langmuir, and Ko-ichi Nakamura, "Hydrothermal Exploration with the Autonomous Benthic Explorer," *Deep Sea Research Part I: Oceanographic Research Papers* 55, no. 2 (February 2008): 203–19.

194 **In 2010 it used similar techniques to map the underwater plume:** Richard Camilli et al., "Tracking Hydrocarbon Plume Transport and Biodegradation at Deepwater Horizon," *Science* 330, no. 6001 (October, 2010): 201–4.

195 **"we had no idea what was happening below":** Author interview with Rich Camilli, Woods Hole, MA, August 2011.

199 **"We want to make cars that are better than drivers":** Burkhard Bilger, "Auto Correct," *The New Yorker*, November 25, 2013, http://www.newyorker.com/ reporting/2013/11/25/131125fa_fact_bilger?currentPage=all.

200 **"without traffic accidents or congestion":** Sebastian Thrun, "Self-Driving Cars Can Save Lives, and Parking Spaces," *New York Times*, December 5, 2011, http:// www.nytimes.com/2011/12/06/science/sebastian-thrun-self-driving-cars-can -save-lives-and-parking-spaces.html. Sebastian Thrun, "What We're Driving At," Google official blog, http://googleblog.blogspot.com/2010/10/what-were-driving -at.html, accessed July 10, 2014. John Markoff, "A Trip in a Self-Driving Car Now Seems Routine," *Bits Blog*, http://bits.blogs.nytimes.com/2014/05/13/a-trip-in-a -self-driving-car-now-seems-routine, accessed July 10, 2014. John Markoff, "Google Cars Drive Themselves, in Traffic," *New York Times*, October 9, 2010, http://www.nytimes.com/2010/10/10/science/10google.html.

200 **The Google car's successful driving tests:** Mark Harris, "How Google's Auto-nomous Car Passed the First U.S. State Self-Driving Test," IEEE Spectrum Online, September 10, 2014, http://spectrum.iee.org. *Idem.*, "These Are the Secrets Google Wanted to Keep about Its Self-Driving Cars," *Quartz*, http://qz .com/252817/these-are-the-secrets-google-wanted-to-keep-about-its-self-driving -cars/, accessed November 18, 2014. Mark Harris, "How Much Training Do You Need to Be a Robocar Test Driver? It Depends On Whom You Work For," IEEE

Spectrum Cars That Think, February 24, 2015, http://spectrum.ieee.org/cars
-that-think/transportation/human-factors/how-much-training-do-you
-need-to-be-a-robocar-test-driver-it-depends-on-whom-you-work-for.

201 **He put a video camera on the dashboard of his car:** John Leonard, "Conversations on Autonomy," presentation, MIT, March 13, 2014. John Markoff, "Police, Pedestrians and the Social Ballet of Merging: The Real Challenges for Self-Driving Cars," *Bits Blog*, http://bits.blogs.nytimes.com/2014/05/29/police-bicycl ists-and-pedestrians-the-real-challenges-for-self-driving-cars/, accessed July 10, 2014.

201 **We know that driverless cars will be susceptible:** John Markoff, "Collision in the Making Between Self-Driving Cars and How the World Works," *New York Times*, January 23, 2012, http://www.nytimes.com/2012/01/24/technology/ googles-autonomous-vehicles-draw-skepticism-at-legal-symposium.html. Will Knight, "Proceed with Caution toward the Self-Driving Car," *MIT Technology Review*, April 16, 2013, http://www.technologyreview.com/review/ 513531/proceed-with-caution-toward-the-self-driving-car/. M. L. Cummings and Jason Ryan, "Shared Authority Concerns in Automated Driving Applications," *Journal of Ergonomics*, S3:001. doi:10.4172/2165-7556.S3-001

202 **how will they rush into the loop quickly enough:** Bianca Bosker, "No One Understands the Scariest, Most Dangerous Part of a Self-Driving Car: Us," *Huffington Post*, September 16, 2013, accessed July 10, 2014.

202 **Google discovered that "people are lazy":** Tom Simonite, "Lazy Humans Shaped Google's New Autonomous Car," *MIT Technology Review* (May 30, 2014), http://www.technologyreview.com/news/527756/lazy-humans-shaped -googles-new-autonomous-car/. Will Knight, "Driverless Cars Are Further Away Than You Think," *MIT Technology Review* (October 22, 2013), http://www .technologyreview.com/featuredstory/520431/driverless-cars-are-further-away -than-you-think/.

203 **"kick back, relax, and enjoy the ride":** Chris Urmson, "Just Press Go: Designing a Self-Driving Vehicle," Google official blog, May 27, 2014, http://googleblog .blogspot.com/2014/05/just-press-go-designing-self-driving.html, accessed July 9, 2014. Evan Ackerman, "Google's Autonomous Cars Are Smarter Than Ever at 700,000 Miles," IEEE Cars that Think Blog, April 29, 2014, accessed July 10, 2014.

205 **published the details of the crash:** Luke Fletcher et al., "The MIT–Cornell Collision and Why It Happened," *Journal of Field Robotics* 25, no. 10 (2008): 775–807.

206 **Lawyers and legal scholars are just beginning:** For example, see David C. Vladeck, "Machines Without Principals: Liability Rules and Artificial

Intelligence," *Washington Law Review* 89, no. 1 (March 2014): 117–50. Curtis Karnow, "The Application of Traditional Tort Theory to Embodied Machine Intelligence," paper presented at the Robotics and Law Conference, Center for Internet and Society, Stanford Law School, April 2013. Also see the blog of the Artificial Intelligence and Robotics Committee of the American Bar Association, http://apps.americanbar.org/dch/committee.cfm?com=ST248008.

206 **The certification approach for software on life-critical systems:** See, for example, DO-178B, the software certification standard required by the FAA: http://en.wikipedia.org/wiki/DO-178B.

209 **define levels of automation in cars:** Erik Stayton, "Driverless Dreams: Narratives, Ideologies, and the Shape of the Automated Car," Thesis, Massachusetts Institute of Technology, 2015.

211 **"Navy Drones with a Mind of Their Own":** James Paduano, et al., "TALOS: An Unmanned Cargo Delivery System for Rotorcraft Landing to Unprepared Sites." Submitted to American Helicopter Society 2015 Annual Forum, May 2015. Dionne Nissenbaum, "Navy Drones with a Mind of Their Own," *Wall Street Journal*, April 5, 2014. For a more nuanced view, see Graham Warwick, "Thinking Helicopters: Manned or Unmanned, Rotorcraft Stand to Benefit from Autonomy," *Aviation Week & Space Technology* (April 21, 2014): 26–7.

213 **Its 2012 pilot report on an aircraft called Centaur:** Fred George, "'Flying' the Centaur Optionally Piloted Aircraft," *Aviation Week & Space Technology* (August 6, 2012).

CHAPTER 7: AUTONOMY IN THE HUMAN WORLD

219 **"I don't use the word 'drone'":** Sara Sorcher, "Drone Lobbyist: 'I Don't Use the Word Drone," *National Journal*, March 27, 2013, http://www.nationaljournal.com/daily/drone-lobbyist-i-don-t-use-the-word-drone-20130327

223 **"The idea of autonomy denies":** Matthew Crawford, *Shop Class as Soulcraft: An Inquiry into the Value of Work* (New York: Penguin Books, 2010).

INDEX